HIGHER EDUCATION AND SDG14

PRAISE FOR *HIGHER EDUCATION AND SDG14*

'[This] is an ambitious series and I am especially interested to know that attention will be paid to SDG14, "Life Below Water", dealing with a vital part of our planet that calls for urgent support. The book in question will clearly make a valuable contribution to our understanding of what needs to be done.'

'As the Series Editor, Professor Purcell points out that higher education is well placed to drive the agenda forward. In my own research and advocacy, I have always emphasised the importance of harnessing the skills, knowledge and energy that young people can bring to the campaign. Delivering this crucial goal for the ocean requires action at all levels, through small organisations as well as large. We cannot simply wait for national governments and international bodies to take a lead.'

James Alix Michel, Former President of the Republic of Seychelles and Chairman of the James Michel Foundation

HIGHER EDUCATION AND THE SUSTAINABLE DEVELOPMENT GOALS

Series Editor

Wendy Purcell
Emeritus Professor and University President Emerita, and Academic Research Scholar with Harvard University

About the Series
Higher Education and the Sustainable Development Goals is a series of 17 books that address each of the SDGs in turn specifically through the lens of higher education. Adopting a solutions-based approach, each book focuses on how higher education is advancing delivery of sustainable development and the United Nations global goals.

Forthcoming Volumes
Higher Education and SDG11: Sustainable Cities and Communities edited by Julio Lumbreras and Jaime Moreno-Serna

Higher Education and SDG2: Zero Hunger edited by Karen Cripps and Pariyarath Sangeetha Thondre

Higher Education and SDG4: Quality Education edited by Tawana Kupe

Higher Education and SDG16: Peace, Justice and Strong Institutions edited by Sarah E. Mendelson

Higher Education and SDG10: Reduced Inequalities edited by Priya Grover, Nidhi Phutela, and Pragya Singh

Higher Education and the Sustainable Development Goals

HIGHER EDUCATION AND SDG14

Life Below Water

EDITED BY

SIMON J. DAVIES
Harper Adams University, UK

AND

PAUL ROBERT VAN DER HEIJDEN
Global Ambassador of Aquaculture without Frontiers, The Netherlands

United Kingdom – North America – Japan – India
Malaysia – China

Emerald Publishing Limited
Emerald Publishing, Floor 5, Northspring, 21-23 Wellington Street, Leeds LS1 4DL.

First edition 2024

British Library Cataloguing in Publication Data
A catalogue record for this book is available from the British Library

ISBN: 978-1-83549-253-6 (Print)
ISBN: 978-1-83549-250-5 (Online)
ISBN: 978-1-83549-252-9 (Epub)

Printed and bound by CPI Group (UK) Ltd, Croydon, CR0 4YY

CONTENTS

PREFACE

Professor Wendy Purcell, PhD FRSA

Higher education (HE) makes an important contribution to the delivery of the Sustainable Development Goals (SDGs). Through high-quality teaching and learning, HE supports the development of responsible citizens as scholars, leaders, entrepreneurs, and professionals. Universities and colleges undertake curiosity-driven and socially impactful research to help advance knowledge frontiers and find solutions for the world's most pressing issues. HE is also active in civic and community settings, often as anchor institutions. Nevertheless, given the fierce urgency of (un)sustainable development, the climate crisis, and widening inequity within countries and across the globe, HE needs to do more and go faster. For HE to deliver fully against the SDGs, it needs to adapt to this shared global agenda and embrace transformative change.

This book series focuses on the role of HE in advancing the SDGs, identifying some successes to date and pointing to opportunities ahead. In sharing the ways and means universities and colleges across the world are engaging with the SDGs, the series seeks to both inspire and enable those in the HE sector and stakeholders beyond to transform what they do and how they do it and hasten progress towards Agenda 2030. Insights gleaned from relevant case studies, innovations, reflective accounts, and student stories can help the HE sector both deepen and accelerate its engagement with the SDGs. Each book seeks to capture examples of how HE institutions (HEIs) are fulfilling the delivery of their academic mission *and* progressing the SDG concerned. Illustrating the work of students, that is undertaken by faculty and staff of the institution and conducted with others, positions HE as a change agent operating at a systems level to help create a world that leaves no one behind.

Taking up this global challenge, SDG14 'Life Below Water' calls on us to 'conserve and sustainably use the oceans, seas, and marine resources for sustainable development', and thereby safeguard the planet's largest ecosystem. Healthy oceans and seas are essential to life, supplying food, energy, water, and supporting employment. However, ocean warming, acidification, over-fishing, algal blooms, biodiversity loss, and plastic pollution represent an ocean emergency and pose a threat to health and wellbeing. HE is at the forefront of science and education to help tackle these issues. Bringing key assets of curiosity and the pursuit of knowledge and its application to partners seeking solutions and driving innovation, universities, and colleges operate in global networks. Helping realise human potential connects the worlds of learning, work, and entrepreneurship in support of more inclusive economic growth. As placemakers, HEIs can use their convening power to draw stakeholders around a problem in support of the adaptive change needed to tackle the challenges of sustainable development.

This book on HE and SDG14 highlights the work of universities and colleges on sustaining life below water with examples drawn from research, knowledge sharing, and the activities of talented students and faculty. Several chapters reflect on the transformational experience of education, fuelling the desire to explore and discover through research and scholarship, and how personal passions related to the sea-shaped lives and careers. Some of the student stories capture the importance of science in action and being involved in projects that have real-world impact, with student mobility a source of innovation. The book includes examples of HE's work on nature-based solutions, sustainable aquaculture, and building with nature. The wonder of our oceans and the fragility of our relationship with life below water comes through the cases and reflects the dynamic tension between conservation and stewardship on the one hand and development and growth on the other.

The health of people, planet, and shared prosperity rely upon the full participation of HE with universities and colleges in turn needing to pursue greater engagement with the SDGs. As organisations that have stood for many centuries in some cases, this demands that they adapt to new models of learning, research partnerships, and leadership and governance frameworks to accelerate progress on delivering the SDGs. Immersive engagement with the SDGs can catalyse pedagogic innovation, serve to refresh curricula, and stimulate new programme development. It can also open new avenues for research, attract new sources of funding and energise people to deliver on the academic mission. Sustainability is a goal for today and sustainable development is an organising principle. HEIs can play a critical role in developing new systemic and transformative solutions through interdisciplinary and multi-stakeholder collaboration and a purposeful focus on the SDGs. This book illustrates this approach as it relates to HE and SDG14.

1

SUSTAINABLE DEVELOPMENT GOAL 14 IN THE 2020s AND BEYOND

Simon J. Davies[a] and
Paul Robert van der Heijden[b]

[a]*Harper Adams University, UK*
[b]*Global Ambassador of Aquaculture Without Frontiers, Netherlands*

ABSTRACT

The chapter provides an overview of the book and addresses the rationale for the selection of cases reflecting teaching and research in major areas of SDG14. For example, the impact of increasing global sea temperature, ocean acidification, and pollution on aquatic life and biosciences. Fisheries and aquaculture for seafood and marine ingredients and marine protected areas (MPAs) that favour the assemblage of fish, crustaceans, alga, coral, and mussels to enhance and stimulate biodiversity. New products derived from marine biotechnology are viewed to conserve and sustainably use the seas and oceans whilst promoting wealth creation and employment. Marine parks allow scientists to better study the marine environment and explore sustainable balances between tourism, work, and recreation in harmony with the Life Below Water – SDG14 mandate. Finally, the aspects of governance and roles of stakeholders and societal involvement are advocated in achieving the safe and effective use of marine resources. Throughout, the role of higher education in providing educated scientists and multidisciplinary specialists for future generations to come is highlighted.

Keywords: SDG14; climate change; ocean acidification; fisheries and aquaculture; pollution; offshore energy; biotechnology; governance; higher education

INTRODUCTION

In the first quarter of the 21st century, we witness global climate threats and pollution events. These consist of toxic chemical agents, plastics, and micro/nano plastics, as well as atmospheric rises in carbon dioxide (CO_2), elevating global temperature and leading towards ocean acidification (Hoegh-Guldberg et al., 2017), both tropical and cold-water corals. Almost half of the world's coral reefs have been lost or severely eroded. Scientists predict that all corals will suffer by 2050 and that 75% could face critical threat levels (Eddy et al., 2021). The extensive impact of overfishing worldwide is a fact, with notable consequences for the food chain and associated biodiversity. The Monaco Blue Foundation (as stated in the Monaco Blue Initiative report 2023) has been instrumental for decades in advocating conservation strategies. Life Below Water has clear anthropogenic connections with the global population expected to reach 9.7 billion by 2050 (UN, 2022). Higher education plays a critical role in addressing many important areas and the following topics serve as examples to be highlighted in successive chapters.

Increased Temperatures

Global climate change associated with rising CO_2 and methane emissions is a high priority. Warming oceans cause significant effects on the life cycle of organisms such as phytoplankton and zooplankton. Disruption of the breeding behaviour of pelagic fish species may extend to invertebrates such as *coelenterates* (jellyfish, corals, and sea anemones) and *mollusks* (e.g. mussels, clams, squid, cuttlefish, octopus). There are also noticeable effects on our fisheries and seafood economy, as described by Hollowed et al. (2013).

Ocean Acidification and Marine Pollution

The 2023 estimate is that nearly 15% of coral reefs have been damaged and impaired over the last decade (UN, 2021). Between 2009 and 2018, the continuous rise in sea temperature has resulted in a loss of 14% of coral reefs – greater than the size of Australia's reefs combined (UN, 2021). There is increased research motivation to include biology and chemistry on the effects of ocean acidification on marine and aquatic fauna. Anthropogenic activities are the primary causes of toxic marine and aquatic pollutants, leading to significant disruption to life below water. Run-off from the land because of herbicides, fertilisers, and industrial chemicals used in the manufacture of products such as paints, pharmaceuticals, flame retardants, (PCB) plasticisers, oils, and heavy metals such as catalysts in the steel industry have been prime areas of concerns (Hatje et al., 2022).

Fisheries and Aquaculture

Aquaculture can provide a source of protein for human consumption, meeting the growing demand for seafood and reducing the pressure on wild fish populations. By providing an alternative source of seafood, aquaculture can help to reduce the demand for wild-caught fish, which can help support sustainable fishing practices and protect marine ecosystems (Naylor et al., 2021). However, marine ingredient allocation in the context of empirical net aquaculture biomass production of fish and shrimp from marine resources (Fish In: Fish Out [FIFO] ratio) was comprehensively evaluated by Kok et al. (2020). Sustainable and responsible fisheries minimise negative impacts on the environment and maximise the social and economic benefits to local communities. Haas et al. (2019) provided an insight into how large-scale commercial fishing operations, with a focus on the harvest industry, can articulate with the seven primary targets of SDG14, providing specific examples for several of the targets (REF to SDG14 targets needed). Four key fisheries areas meeting SDG14 targets are presented in Table 1.

Offshore Energy Generation

Wave energy systems are typically located away from sensitive coastal habitats and wind farms can be sited in areas with low biodiversity to minimise their impact, thus reducing the need for coastal development and habitat destruction. The progress and achievements of wave energy development in the UK towards achieving net zero for the UK industry by 2030 are reviewed and described by Jin and Greaves (2021). This may stimulate local economies and support sustainable development in coastal communities. Wind farms are good for fisheries as they will create no-take areas that may work as pseudo-protected areas and favour the aggregation of organisms like algae, corals, and mussels, creating a suitable environment that attracts fish (Michler-Cieluch and Kodeih, 2008).

Table 1. The Sustainable Fisheries Agenda for Meeting SDG14.

Protecting marine ecosystems	By adopting sustainable fishing practices, such as using selective fishing gear and setting catch limits, responsible fisheries can help to protect marine ecosystems and preserve the diversity of marine life.
Responsible fisheries	Responsible fisheries can provide a sustainable source of food for human consumption, helping to meet the growing demand for seafood in a way that does not harm the environment.
Providing economic benefits	Responsible fisheries can benefit local communities economically, particularly in coastal areas where fishing is an essential source of livelihood.
Promoting sustainable development	By adopting sustainable and responsible practices, fisheries can contribute to the overall goal of SDG14 to conserve and sustainably use marine resources for sustainable development.

Importance of Marine Biotechnology

Marine biotechnology refers to using marine resources, such as microorganisms, plants, animals, and other marine-derived natural materials, to develop products and technologies that benefit society, as recently defined by Rotter et al. (2021). Marine biotechnology has the potential to support SDG14, which aims to conserve and sustainably use the oceans for sustainable development.

Marine Protected Areas

Marine parks are areas of the ocean set aside for protecting and conserving marine life and ecosystems, which was described in scientific terms as a primary mechanism for the guardianship and restoration of marine ecosystems by Turnbull et al. (2021). As Davies (2023) stated recently, they can play crucial roles in the conservation of the environment, including protecting a wide range of marine species and habitats, helping to stimulate and enhance the diversity of life in the ocean. Marine parks serve as a place for scientists to research marine life and ecosystems, which helps inform conservation efforts and can benefit local communities economically through tourism and recreation.

Governance and Societal Involvement

Effective governance is essential for coordinating and aligning the efforts of various stakeholders, including governments, businesses, non-governmental organisations (NGOs), and local communities, to achieve SDG14. This can involve setting targets and indicators to measure progress, and providing resources and support for implementation, as explained by Haward and Haas (2021). It is essential to characterise projected changes in the scale and size of the ocean economy and the part that observations, measurements, and forecasts contribute towards supporting the safe, effective, and equitable use of the ocean's resources (Rayner et al., 2019). Sumaila et al. (2021) provided a strategic assessment of financing a sustainable ocean economy.

THE ROLE OF HIGHER EDUCATION

Higher education institutions can offer courses that focus specifically on marine biology, oceanography, and other disciplines related to studying Life Below Water. They can also support and conduct research on issues related to SDG14, such as the impacts of climate change on marine ecosystems or the management and conservation of marine resources. These programmes can provide students with the knowledge and skills needed to understand the complex relationships between humans and the ocean whilst appreciating the challenges and opportunities ahead.

Here, the role of higher education in helping to meet the targets of SDG14 is presented as case studies that show how the sector provides opportunities and enrichment for students and researchers to highlight their originality, enterprise, and innovation. We have endeavoured in our selection to present interesting and timely work in leading universities and colleges that can be amplified by others. These examples reflect the current contribution of universities and colleges to SDG14 and signal where more progress can be made. The chapters are carefully sequenced to provide a logical flow from the riverine to coastal areas transitioning to our seas and ocean environment providing a comprehensive journey.

We commence with an assessment of water quality criteria within a river system in Southwest England, with significance to aquatic life and biodiversity. Vera Cirina, a student in the UK evaluates anthropogenic activities with an emphasis on the catchment of Nitrogen and metals on water quality by efficient monitoring to build a model. Her report includes recommendations and mitigation strategies needed to improve the environmental quality of the system.

Matt Bell, a Marine Biology and Oceanography student from the UK describes his work on surveying the local marine biodiversity in the possible preparation of the installation of a subsea tidal generator in Southwest England. He depicts how underwater marine engineering for renewable energy schemes served as the basis for a full diving and technical experience, as part of his assignment to assess various potential factors and criteria that may affect the benthos fauna.

Martin Baptist from the Netherlands examines challenges in safeguarding its low-lying coastline against rising sea levels and the consequences of coastal defence strategies on marine life, particularly in relation to SDG14. The use of hard structures for coastal protection contributes to the loss of natural coastal habitats, raising biodiversity concerns.

Caterina Pezzola from Italy describes the vital importance of seaweeds in marine ecology and algal bloom issues in Tuscany. Re-purposing the biomass to produce seaweed-derived commercial goods would provide benefits for the environment and local economic activities while promoting a sustainable business within a circular bio-economy framework.

Dominic Duncan Mensah (PhD student) from Ghana and with colleagues from his university in Norway explored a group of highly promising emerging novel ingredients known as microbial ingredients (MIs), means of producing them, and how they can help achieve sustainable aquaculture and SDG14 targets. This chapter interestingly connects Life on Land SDG15 with SDG14 showing overlap and synergy offering novel solutions and bioconomical considerations.

Richard Page, an academic from the USA based in the Pacific Island State of Palau and college student Chris de Blok present a case study of Palau, one of the first countries to use marine protected areas (MPAs) as a tool to develop biodiversity within its Exclusive Economic Zone (EEZ). Pacific MPAs impact on ecosystems and fisheries are compared with European MPAs, with examples such as wind turbine arrays, artificial reefs, and ecosystem and fisheries impacts.

An overview of Sustainable Development Goal 14 in relation to aquaculture and fisheries is given by American, James Sibley from Boston, Massachusetts with contributions from Matt Bell and Simon J. Davies. They address feed from sustainable sources and highlight advances in marine biotechnology and molecular genomics to produce healthy viable farmed ocean fish like Atlantic salmon. Aquaculture is a major seafood contributor and Sibley brings this into focus.

Davy van Doren of MatureDevelopment (Voorburg, the Netherlands) provides an insight into the higher education basis for his work on Building with Nature – an initiative towards sustainable hydraulic engineering. This chapter assesses how issues in relation to nature conservation and social wellbeing are being addressed in practice and their effectiveness on voluntary green behavioural and operational changes, particularly in relation to challenges and benefits associated to the enforcement of corporate responsibility. This was mainly a case study situated in Abu Dhabi, and the construction of Khalifa Port, UAE.

Finally, Richard Pompoes a doctoral student from Germany and the Netherlands examines how diverse actors in the Cauvery Delta, India, and the Mekong Delta, Vietnam, understand and live with water salinity. This chapter addresses some of the SDG14 targets, i.e. SDG14.2 ('Protect and restore ecosystems') and 14.5 ('Conserve coastal and marine areas') with changing delta profiles and the effects on local stakeholders and societal involvement. It includes how his research articulates teaching and student interactions and learning outcomes integral to SDG14.

We can be optimistic for the future of how we can preserve and protect our aquatic and marine bioresources via our younger educators and scientists and their efforts to pass on this message to future generations. This book and its constituent chapters not only give prominence to a selection of topics but also serve as a platform for successive generations in Higher Education to lead and uphold such values as stewardship, ownership, and entrepreneurship to value our planet's unique water ecosystem.

REFERENCES

Davies, S.J. 2023. Utilizing marine protected areas to support aquaculture operations, *International Aquafeed*, September Issue.

Eddy, T.D., Lam, V.W.Y., Reygondeau, G., Cisneros-Montemayor, A.M., Greer, K., Palomares, M.L.D., Bruno, J.F., Ota, Y. and Cheung, W.W.L. 2021. Global decline in capacity of coral reefs to provide ecosystem services, *One Earth*, 4(9), 1278–1285, ISSN 2590-3322. doi: 10.1016/j.oneear.2021.08.016

Haas, B., Fleming, A., Haward, M. and McGee, J. 2019. Big fishing: the role of the large-scale commercial fishing industry in achieving

Sustainable Development Goal 14, *Reviews in Fish Biology and Fisheries*, 29(1), 161–175. doi: 10.1007/s11160-018-09546-8

Hatje, V., Sarin, M., Sander, S.G., Omanović, D., Ramachandran, P., Völker, C., Barra, R.O. and Tagliabue, A. 2022. Emergent interactive effects of climate change and contaminants in coastal and ocean ecosystems, *Frontiers in Marine Science*, 9, 936109. doi: 10.3389/fmars.2022.936109

Haward, M. and Haas, B. 2021. The need for social considerations in SDG 14, *Frontiers in Marine Science*, 8. doi: 10.3389/fmars.2021.632282

Hoegh-Guldberg, O., Poloczanska, E.S., Skirving, W. and Dove, S. 2017. Coral reef ecosystems under climate change and ocean acidification, *Frontiers in Marine Science*, 4(MAY). doi: 10.3389/fmars.2017.00158

Hollowed, A.B., Barange, M., Beamish, R.J., Brander, K., Cochrane, K., Drinkwater, K., Foreman, M.G.G., Hare, J.A., Holt, J., Ito, S., Kim, S., King, J.R., Loeng, H., MacKenzie, B.R., Mueter, F.J., Okey, T.A., Peck, M.A., Radchenko, V.I., Rice, J.C., Schirripa, M.J., Yatsu, A. and Yamanaka, Y. 2013, September. Projected impacts of climate change on marine fish and fisheries, *ICES Journal of Marine Science*, 70(5), 1023–1037. doi: 10.1093/icesjms/fst081

Jin, S. and Greaves, D. 2021. Wave energy in the UK: status review and future perspectives, *Renewable and Sustainable Energy Reviews*, 143, 110932. doi: 10.1016/j.rser.2021.110932

Kok, B., Malcorps, W., Tlusty, M.F., Eltholth, M.M., Auchterlonie, N.A., Little, D.C., Harmsen, R., Newton, R.W. and Davies, S.J. 2020. Fish as feed: using economic allocation to quantify the Fish in - Fish-out ratio of major fed aquaculture species, *Aquaculture*, 528. doi: 10.1016/j.aquaculture.2020.735474

Michler-Cieluch, T. and Kodeih, S. 2008. Mussel and seaweed cultivation in offshore wind farms: an opinion survey, *Coastal Management*, 36(4), 392–411. doi: 10.1080/08920750802273185

Monaco Blue Initiative, Earth Negotiations Bulletin. 2023. (ENB) International Insitute for Sustainable Development 23rd March 2023, 1–5.

Naylor, R.L., Hardy, R.W., Buschmann, A.H., Bush, S.R., Cao, L., Klinger, D.H., Little, D.C., Lubchenco, J., Shumway, S.E. and Troell, M. 2021.

A 20-year retrospective review of global aquaculture, *Nature*, 591(7851), 551–563. doi: 10.1038/s41586-021-03308-6

Rayner, R., Jolly, C. and Gouldman, C. 2019. Ocean observing and the blue economy, *Frontiers in Marine Science*. doi: 10.3389/fmars.2019.00330

Rotter, A., Barbier, M., Bertoni, F., Bones, A.M., Cancela, M.L., Carlsson, J., Carvalho, M.F., Cegłowska, M., Chirivella-Martorell, J., Conk Dalay, M., Cueto, M., Dailianis, T., Deniz, I., Díaz-Marrero, A.R., Drakulovic, D., Dubnika, A., Edwards, C., Einarsson, H., Erdoğan, A., Eroldoğan, O.T., Ezra, D., Fazi, S., FitzGerald, R.J., Gargan, L.M., Gaudêncio, S.P., Udovič, M.G., DeNardis, N.I., Jónsdóttir, R., Kataržyte, M., Klun, K., Kotta, J., Ktari, L., Ljubešić, Z., Bilela, L.L., Mandalakis, M., Massa-Gallucci, A., Matijošyte˙, I., Mazur-Marzec, H., Mehiri, M., Nielsen, S.L., Novoveská, L., Overlinge˙, D., Perale, G., Ramasamy, P., Rebours, C., Reinsch, T., Reyes, F., Rinkevich, B., Robbens, J., Röttinger, E., Rudovica, V., Sabotič, J., Safarik, I., Talve, S., Tasdemir, D., Schneider, X.T., Thomas, O.P., Toruńska-Sitarz, A., Varese, G.C., Vasquez, M.I. 2021. The essentials of marine biotechnology, *Frontiers in Marine Science*, 8, doi: 10.3389/fmars.2021.629629

Sumaila, U.R., Walsh, M., Hoareau, K., Cox, A., Teh, L., Abdallah, P., Akpalu, W., Anna, Z., Benzaken, D., Crona, B., Fitzgerald, T., Heaps, L., Issifu, I., Karousakis, K., Lange, G.M., Leland, A., Miller, D., Sack, K., Shahnaz, D., Thiele, T., Vestergaard, N., Yagi, N. and Zhang, J. 2021. Financing a sustainable ocean economy, *Nature Communications*, 12(1). doi: 10.1038/s41467-021-23168-y

Turnbull, J.W., Johnston, E.L. and Clark, G.F. 2021. Evaluating the social and ecological effectiveness of partially protected marine areas, *Conservation Biology*, 35(3), 921–932. doi: 10.1111/cobi.13677

United Nations. 2021. UN News, Global Perspectives Human Stories; Decade of Climate Breakdown Saw 14 Percent of Coral Reefs Vanish, Climate and Environment, October 5. Available at: https://news.un.org/en/story/2021/10/1102242

United Nations. 2022. *UN News, Global Perspective Human Stories*. Available at: https://news.un.org/en/story/2022/07/1122272

2

ASSESSMENT OF RIVER QUALITY FOR AQUATIC LIFE: A SOUTH ENGLAND CASE STUDY

Vera Cirinà

University of Plymouth, UK

ABSTRACT

The report is an environmental impact assessment of two conjoining water streams in the lower area of the Wembury catchment where freshwater meets the coast. The assessment was conducted as there were concerns that the streams may be causing exceedances of the Environmental Quality Standards (EQS) due to catchment-based inputs from anthropogenic activities.

Water, sediment samples, and other parameters were collected, measured, and treated to preserve the concentration of nutrients and metals at three sites. A comparison with the neighbouring river Erme was made to determine if the findings were coherent. The report includes recommendations and mitigation strategies needed to improve the environmental quality of the system.

Findings indicate several breaches of EQS: water nitrogen, copper and zinc, and sediment copper. The highest recorded concentrations were mainly at sites one and two, likely from point source inputs from Wembury town and pollution accumulation from upstream land use such as arable and agricultural land. A special precaution must be taken for sediment copper, increasing monitoring to ensure values do not exceed Probable Effects Level (PEL) possibly becoming dangerous for the fauna and flora but also for humans. River Erme showed to also have EQS breaches but to some degree displayed an overall better ecological status. Despite several breaches in the legal limits, Wembury displays an overall good ecological status supporting life above and below water and is therefore an appropriate model for promoting environmental stewardship. It is to be noted that the material of the coursework was further edited after its original submission.

Keywords: Water quality; Environmental Quality Standards; nutrient contamination; heavy metal contamination; mitigating human impacts

HIGHER EDUCATION

The University of Plymouth offers a cutting-edge academic side with modern facilities, and a great focus is placed on sustainability and the impacts and mitigations of climate change. Linking the latter to Sustainable Development Goal (SDG) 14, the author had the opportunity to study world fisheries and global stocks, aquaculture and endangered species, environmental law, including maritime and ocean law. All her modules explicitly incorporate the SDGs. Plymouth University has opened the doors to some unique opportunities such as undertaking the HSE Scuba course (United Kingdom first-level commercial diving qualification), important modules such as Scientific Diving, Managing Human Impacts on Marine Ecosystems, and gaining highly relevant field and research skills both in terrestrial and marine ambient. It has further opened the opportunity for a placement year at Shoreline, the company managing the Miramare Marine Protected Area in Trieste, Italy. Such experiences increase employability, allow one to learn many valuable skills, and give insight into the working world.

The following section is a case study taken from the 'Environmental Field and Research Skills' module, ENVS2001. It is a key example of how anthropogenic activities have affected freshwater bodies and how to improve the water quality and in turn, life below water. This chapter focuses on SDG14, specifically targets 14.1 (reduction of nutrient pollution), 14.2 (sustainable management of coastal ecosystems), 14.3 (ocean acidification), and 14.a (increasing scientific knowledge). The author believes such case studies to be very important not only for the formation of students as pioneers and stakeholders of the future but also for learning and developing standardised methodologies that can be used in very different environments. It is to be noted that the material of the coursework was further edited after its original submission.

WEMBURY REPORT: A CASE STUDY IN SOUTHWEST ENGLAND

Introduction

The preservation of water quality is a fundamental aspect of the protection of rivers, lakes, the ocean, and all water bodies. Water quality is essential for agriculture, both for crop irrigation and livestock, and humans. Water quality on a national scale is linked to SDG14, 3 (Good Health and Well-being), 6 (Clean Water and Sanitation), 12 (Responsible Consumption and Production), 13 (Climate Action), and 15 (Life on Land). On a global scale, especially in coastal communities more dependent on fishing, further SDGs such as 1 (No Poverty), 2 (Zero Hunger), and 8 (Decent Work and Economic Growth) are involved. All SDGs are in fact extremely linked and intertwined while at the same time complementing each other.

Wembury is situated in Southwest Devon, on the outskirts of Plymouth. The area at and around Wembury is of particular beauty and carries historic significance as it was a Royal Navy's chief gunnery school in the middle twentieth century. This area is protected and managed as a South Devon Area of Outstanding Natural Beauty (AONB) and is also defined as a Site of Specific Scientific Interest (SSSI) and a Special Area of Conservation (SAC). The protection of Wembury's coastal habitat aims at target 14.2 (sustainable management of coastal ecosystems). Several organisations, volunteers, and locals work together to maintain the land to a good ecological standard.

The area has also great fauna and flora diversity, inhabiting some rare species, for instance, the self-seeding Plymouth Pear tree (*Pyrus cordata*), and the tortoise beetle (subfamily Cassidinae).

The Wembury report is an environmental impact assessment of two conjoining water streams in the lower area of the catchment where freshwater meets the coast (Fig. 1). The assessment, written from the perspective of an Environmental Agency officer, was conducted due to increased anthropogenic activity in the area which raised concerns about potential breaches in Environmental Quality Standards (EQS). Samples of sediment and water were taken at three locations and specially treated to observe nutrient and metal concentrations. Additionally, to understand the conditions in the area other parameters such as conductivity, dissolved oxygen (DO), and pH were measured. Data manipulation and statistical analysis were conducted on the final data with the task of producing an environmental assessment of the Wembury streams together with a comparison to a neighbouring river to confirm coherence in the findings. Above all, recommendations and mitigation strategies to improve the environmental quality of the system were developed.

Catchment Fieldwork Sites Map

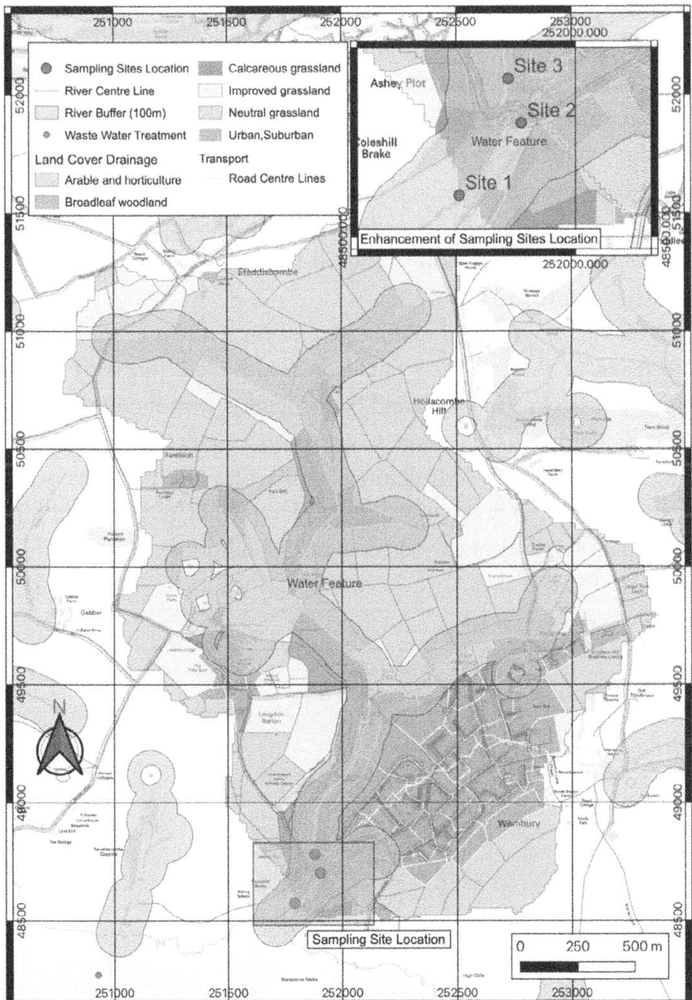

Fig. 1. Map Created Using QGIS Version 3.22.1 Białowieża, Representing the Wembury Catchment with Its Land Uses, the Two Water Streams Under Investigation – River Centre Line Includes a 100-metre Buffer Zone, and the Three Fieldwork Sampling Sites. An Inset Map Highlights the Sampling Sites (Cirina, 2021).

Results and Discussion

The trends of nitrogen and phosphorus concentrations (Fig. 2) both increased between sampling sites one and two, nitrogen rose by 1 (mg•L^{-1}) and phosphorus by 3 (mg•L^{-1}). Nitrogen reached its highest value of 5.4 (mg•L^{-1}) at site two then decreased slightly. Nitrogen at all sites was in excellent correspondence with EQS also when considering the ±2 s.d. On the other hand, all phosphorus concentrations (Fig. 3) exceeded EQS. However, when considering the lower values of the Standard Deviation (SD), only site one breached the EQS.

DO (Fig. 2) remained relatively constant with an overall sample site mean of 9.7 (mg•L^{-1}) but decreased by 1.2 (mg•L^{-1}) at site three. The decline in DO could be associated with increased nutrient concentrations. Pearson Correlation Coefficient (r = −0.74) indicated a negative correlation, supporting the previous statement. Despite the fluctuations in DO levels, they remained in high percentages, indicating good ecological conditions.

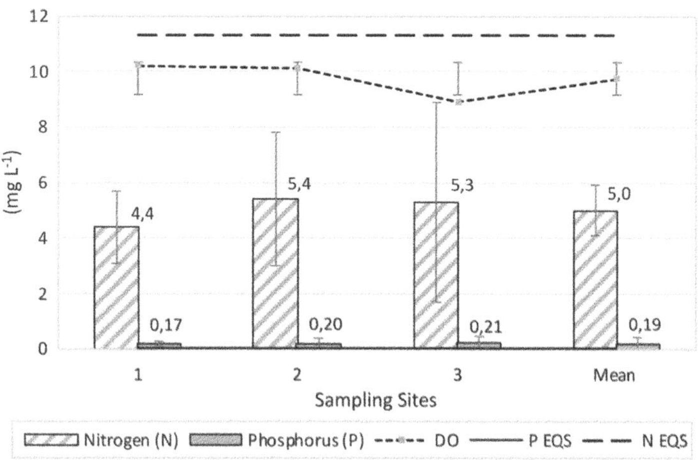

Fig. 2. Mean Nitrogen (N) and Phosphorus (P) (mg•L^{-1}) from Water Samples Collected at the Three Sampling Sites. Dissolved Oxygen (DO) Is Represented by the Blue Line (mg•L^{-1}). EQS of N (11.30 mg•L^{-1}) and P (0.036 mg•L^{-1}) Are Shown as the Yellow and Orange Horizontal Lines. All Represented Data Except EQS Have ± 2 x s.d. Shown as Error Bars.

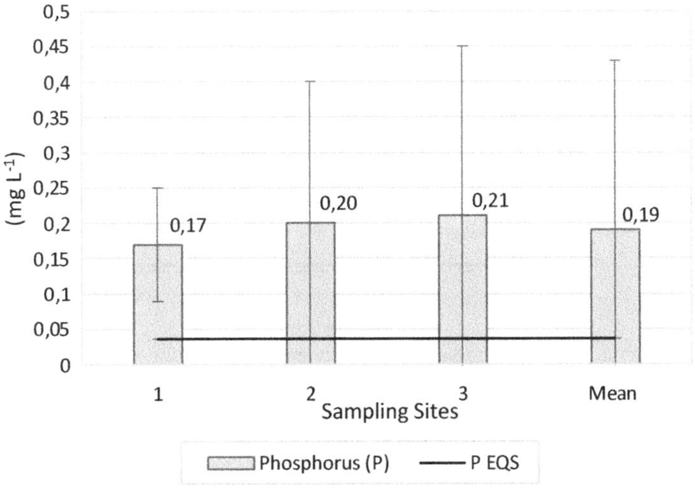

Fig. 3. Enlargement of Fig. 2 for Mean Phosphorus (P) Concentrations (mg•L^{-1}) with ± 2 x s.d. Shown as Error Bars for All Sampling Sites from the Collected Water Sample with EQS (0.036 mg•L^{-1}).

Copper and zinc concentrations (Fig. 4) both exceeded the EQS, this was also the case when considering their SD. Both metals were highest at site one, copper at 0.0409 (mg•L^{-1}) and zinc at 0.03636 (mg•L^{-1}).

However, the Dixons Q-test indicated this copper value to be an outlier. Conductivity (Fig. 4) had an overall site mean of 321 (µS). It experienced a 65 (µS) decrease between sites one and two, then increased again, reaching 316 (µS) at site three. Changes in conductivity can be associated with metal concentrations. Pearson Correlation Coefficient (r = 0.94) indicated a strong positive correlation, supporting the statement, indicating the fluctuations of the two parameters were connected.

The copper concentrations (Fig. 5) at all sites greatly exceeded EQS, with the highest values at site two, 132 (mg•kg^{-1}) and an overall sampling site mean of 113 (mg•kg^{-1}). Also, when considering the SD, concentrations were well above the recommended limits, with possible variations up to 191.8 (mg•kg^{-1}). The SD at site two showed only an 8 (mg•kg^{-1}) difference from the Probable Effect Level (PEL) (Table 1). The PEL is provided to evaluate

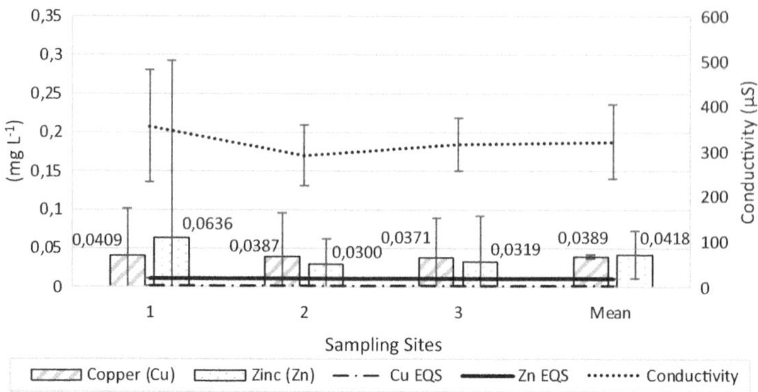

Fig. 4. Mean Copper (Cu) and Zinc (Zn) Concentrations (mg•L^{-1}) from Water Samples Collected at the Three Sites. EQS of Cu (0.028 mg•L^{-1}) and Zn (0.125 mg•L^{-1}) Are Shown as the Yellow and Orange Horizontal Lines. Conductivity (µS) Is Displayed by the Orange Line. All Represented Data Except EQS Have ± 2 x s.d. Shown as Error Bars.

the degree to which adverse biological effects are likely to occur (Environment, 1999c). Differently from copper, zinc concentrations (Fig. 5) were for the majority compliant with EQS, with

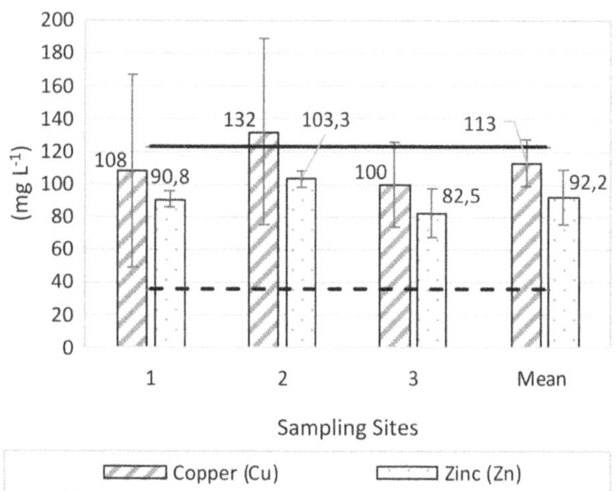

Fig. 5. Mean Copper (Cu) and Zinc (Zn) Concentrations (mg•kg^{-1}) from the Sediment Samples at the Three Locations with ± 2 x s.d. Shown as Error Bars. EQS of Cu (35.7 mg•kg^{-1}) and Zn (123 mg•kg^{-1}) Are Shown as Horizontal Lines.

Table 1. Comparison of Two Water Bodies in SW England: Erme and Wembury Rivers.

Water Data				
	N	**P**	**Cu**	**Zn**
Erme	1.9	0.285	0.0001495	0.0135
Wembury	5.0	0.19	0.0389	0.0418
EQS	**11.30**	**0.036** (0.027–0.050)	**0.001**	**0.011**

Sediment Data		
	Cu	**Zn**
Erme	17–24	125
Wembury	113	92.2
EQS	**35.7**	**123**
PEL	197	315

Notes: Mean nitrogen (N), phosphorus (P), copper (Cu), and zinc (Zn) concentrations from water samples displayed ($mg \cdot L^{-1}$). Mean copper (Cu) and zinc (Zn) concentrations from sediment data are displayed ($mg \cdot kg^{-1}$). Most recent EQS and PEL: (UKTAG, 2013) (DEFRA, Environmental Agency, 2019) (Canadian Council of Ministers of the Environment, 1999), and PEL (Probable Effects Level) (Canadian Council of Ministers of the Environment, 1999). River Erme data: (DEFRA, Environmental Agency, 2021) (British Geological Survey, 2020a, 2020b). N.B. River Erme water data was calculated using the two most recently available measurements.

only site two exceeding the limits. When considering the SD, all sites were in exceedance. Noticeably, both metal concentrations were highest at site two. Phosphorus values (Fig. 6) at all sites were well below EQS including the SD.

Environmental Significance of Water Data

This section provides context to better understand what rising nutrient and metal concentrations in the environment signify. In many parts, SDG14 target 14.1 is pursued, raising attention to the consequences of nutrient pollution, including ocean acidification (target 14.3). Nutrients and trace metals are naturally present in the environment, but when these exceed average background concentrations, they can become dangerous for the environment and its organisms, including humans.

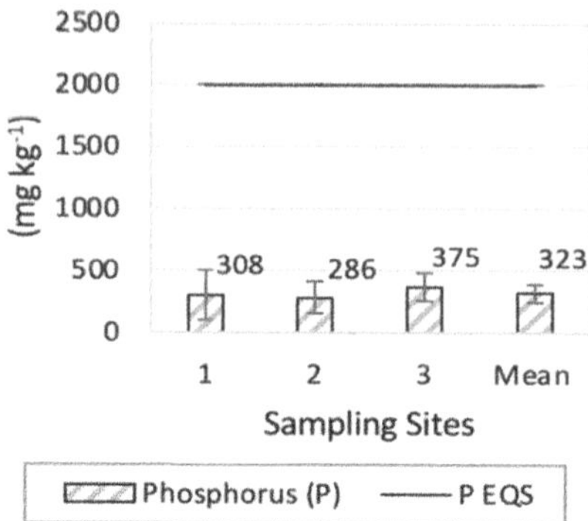

Fig. 6. Mean Phosphorus (P) Concentrations from Collected Sediment Sample (mg•kg^{-1}) with ± 2 x s.d. Shown as Error Bars with Corresponding EQS of 2000 (mg•kg^{-1}).

To understand why concentrations are fluctuating, it is important to know what the main land uses in the Wembury catchment are. The main land uses are arable and horticulture, improved, natural, and calcareous grassland, broadleaf woodland, and urban and suburban (Fig. 1). The northwest catchment area (Site 3) is predominantly influenced by agricultural practices, specifically livestock waste and fertiliser run-off. Site two has agricultural influences but mostly urban ones from Wembury village and the north-east stream which includes a long-abandoned, uninfluential antimony mine upstream (Fig. 7). The two streams converge at site one and flow further into the sea.

The predominant anthropogenic nutrient inputs in the catchment are from agricultural practice, septic discharges, and some atmospheric deposition (Environment, 1999c). Additional inputs could include fertiliser runoff from gardens in Wembury village and wooden fence treatments. Precipitation from storms can also increase nutrient contamination (DEFRA, Environmental Agency, 2021), washing away pollutants from roads and rooftops. Sewage and septic systems are a relevant source of nutrient inputs, particularly phosphorus. Nitrogen has a more water-soluble behaviour and eventually moves from land out to coastal waters, it enters the water body through the oxidation of nitrite, ammonia, and organic nitrogen compounds.

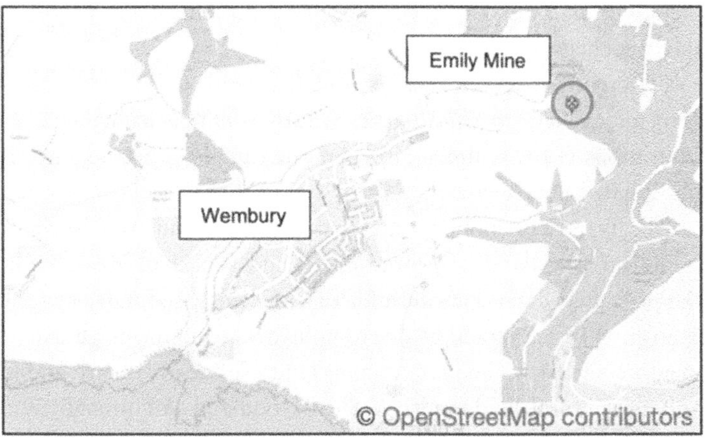

Fig. 7. Map showing mine Emily (also known as Wheal Emily) in proximity of the investigated site. The mine was last operational in 1852. (adapted from Mindat.org, 2022).

The progressive enrichment of nutrient concentrations can boost plant growth potentially leading to eutrophication (Rieuwerts, 2015). Eutrophication causes a chain reaction in the ecosystems, causing an initial overabundance of algae and plants. Excess vegetation eventually decomposes and produces large amounts of carbon dioxide. This can cause pH levels to decrease, contributing to a process known as ocean and estuarine acidification (NOAA, n.d.B). Increased nutrient concentrations also reduce DO levels, possibly leading to hypoxic or anoxic conditions (Environment, 1999c), as well as marine dead zones. Furthermore, DO can be depleted by the biodegradation of respirable organic matter (ROM), when organic matter is degraded by aerobic bacteria this will also contribute to a waterbody's DO depletion (Rieuwerts, 2015). Another consequence includes Harmful Algal Blooms (HAB), which is when a colony of algae grow uncontrollably and produce toxic or harmful effects to aquatic life and in rare cases also to humans (NOAA, n.d.A). These blooms may last up to several months and can cover several kilometres (NOAA, n.d.A). In marine systems, severe cases of over-enriched waters can lead to decreasing (coastal) silicate levels, resulting in numerous changes to local biota (Environment, 1999c). This has important significance to those in the aquatic ecosystem due to the dependence of many marine microorganisms including diatoms that have a specific requirement for silicon in their biochemistry.

Differently from nitrogen, phosphorus tends to be more adsorbent and adhere to solid particles, thus resulting in higher concentrations in the sediment, it enters the catchment in the form of phosphate salts (Rieuwerts, 2015).

The first two sampling sites have similar land uses, such as broadleaf woodland, improved grass, and suburban areas. From the presented data, one can infer that the increased nutrient concentration (Fig. 2) could be due to influences from the upper catchment on site two. The predominant land uses upstream are arable and agricultural land, and broadleaf woodland. Additionally, the presence of urban influences at all three sites was a source of excess nitrate and phosphate.

Metals are naturally present in the environment, but at elevated concentrations can be toxic to aquatic biota (Environment, 1999c). Copper and zinc behave similarly, entering systems through surface runoff and aerial deposition (Environment, 1999c), this also includes domestic sources (e.g. wastewater treatment works), industrial effluents, and mine waters (UK Technical Advisory Group on the Water Framework Directive (UKTAG, 2013a, 2013b). Additional inputs may include land use of galvanised metal (mainly Zn), paints (mainly Zn), garden fence treatments (mainly Cu), and village infrastructure, i.e. roads and concrete stream culverts.

Similarly, to phosphorus, these metals are sparingly soluble and have a strong affinity for particle matter, therefore accumulating more in sediments (Environment, 1999a, 1999b). They are generally found in the form of salts that dissolve in water (Rieuwerts, 2015). Elevated copper concentrations in the environment are of global concern due to its non-biodegradability and potential for accumulation in body tissues – it adversely affects ecosystems and human health (Elkhatat, et al., 2021).

The metal trends (Fig. 4) could be associated with the land use of the sampling sites. One can infer that the higher concentrations at site one could be influenced by the nearby urban area and the roads leading to Wembury Beach. Urban influences also may have impacted sampling points two and three, however, these sites were mainly affected by the upstream of the catchment and the presence of arable and agricultural land.

Environmental Significance of Sediment Data

Sediments act as an important route of exposure to aquatic life as many organisms live in or are in close contact with bed sediment. Adverse biological effects of copper and zinc include decreased benthic invertebrate diversity and abundance, increased mortality, and behavioural changes (Environment, 1999a, 1999b).

Conditions in the sediment samples seemed of a better ecological standard than in the water samples. Phosphorus was in excellent condition with all values well below legal limits, zinc only exceeded the EQS at site two. Copper, on the other hand, breached EQS at all sites. It is important to note, that when considering the copper PEL (Table 1), site two was only 65 ($mg•kg^{-1}$) away. If one considered the SD at that site, it was only 8 ($mg•kg^{-1}$) difference from PEL. If conditions deteriorate, it could cause adverse biological effects on the environment and its biota. The land uses and anthropogenic activities previously mentioned ('Environmental Significance of Water Data' section) could be linked to the trends observed (Figs. 5 and 6). Both metal concentrations were highest at site two, this could be associated with the greater urban influences from Wembury village. It could be observed that between sites one and two, the metal concentrations (Fig. 5) increased, and the phosphorus levels (Fig. 6) decreased. Academic literature suggests that in soil particles, e.g. crop production, zinc absorption capacity is reduced by high phosphorus utilisation, and zinc acts as a natural regulator for phosphorus (Mousavi, 2011). A Pearson Correlation Coefficient ($r = -0.92$) displayed a strong negative correlation, indicating that zinc and phosphorus have an antagonistic state to one another.

Comparison with River Erme

A comparison with river Erme was conducted to assess whether Wembury concentrations are typical for the area (Fig. 8).

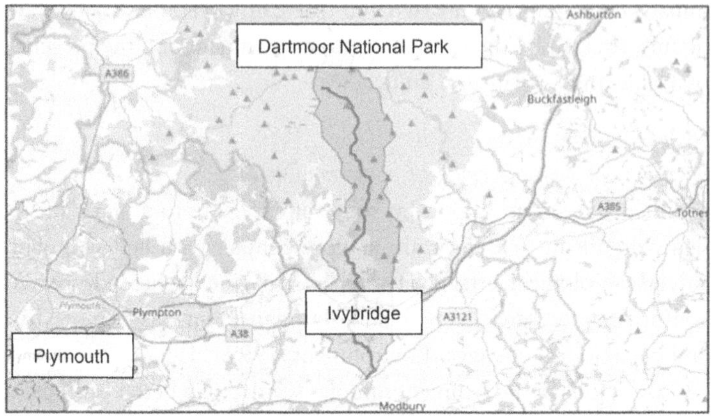

Fig. 8. Map indicating position of Erme Water Body. Located in SW England, 16-km East of Wembury. Beginning in Dartmoor National Park and Flowing through Ivybridge into the South Hams Countryside (DEFRA, Environmental Agency, 2021).

Considering the data from the water samples, Wembury had higher nitrogen, copper, and zinc values. Excluding nitrogen, concentrations at Wembury exceeded EQS. Furthermore, river Emre's phosphorous value exceeded the recommended limit almost by a factor of eight. The water samples of the two rivers had relatively similar concentrations, with the exception of nitrogen and copper which were greater in Wembury. In the sediment samples, it was evident that Wembury's copper concentrations were particularly high, about three times greater than the EQS, while in Erme, levels were relatively low. As previously mentioned in the report, copper concentrations in Wembury were currently not of extreme concern but were moving closer to the PEL. Oppositely, zinc levels in Wembury complied while Erme breached the EQS. Overall, river Erme had a better ecological status than Wembury; it was nevertheless classified by DEFRA (United Kingdom Department for Environment, Food and Rural Affairs) as having moderate ecological status (DEFRA, Environmental Agency, 2021).

Catchment Management Policies and Practical Mitigation Strategies

Analysis of Extra Data Set

An analysis of an additional data set of the Wembury stream was analysed. Both nitrogen and phosphorus (Figs. 9 and 10) exceeded their EQS. Much higher concentrations of nitrogen were observed compared to phosphorus, possibly due to its greater water-soluble characteristic. Similarly, both copper and zinc (Fig. 11) were above their EQS. Zinc was greatly over the recommended limit, with the highest values at site two (0.05 mg•L^{-1}). Copper also had the highest concentrations at site two (0.020 mg•L^{-1}).

Catchment Management Policies and Practical Mitigation Strategies

This section presents current policies and strategies in place to manage and mitigate issues impacting the water quality of the South West River Basin District (SWRBD). The protective system in place

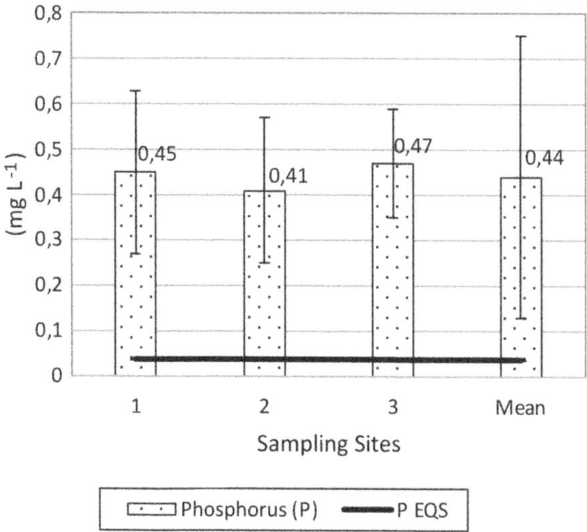

Fig. 9. Freshwater Phosphorus (P) Concentrations (mg•L^{-1}) for All Sampling Locations with ± 2 x s.d. and EQS (0.036 mg•L^{-1}).

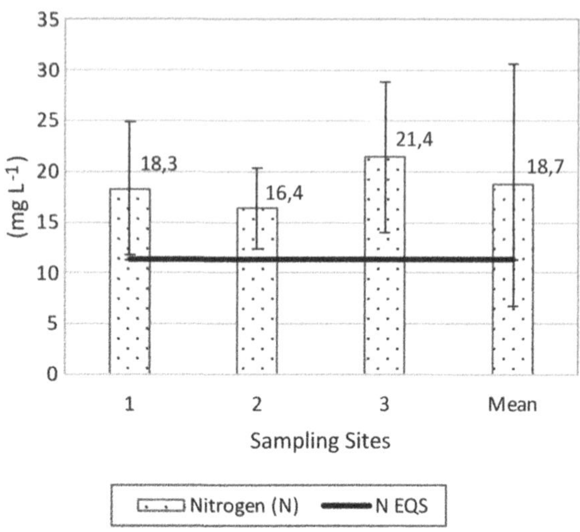

Fig. 10. Mean Nitrogen (N) Concentrations (mg•L⁻¹) from Water Samples at Three Sites with ± 2 x s.d. Shown as Error Bars and EQS (11.30 mg•L⁻¹).

is associated with SDG14 target 14.2 (sustainable management of coastal ecosystems), as it does not only represent the welfare of inland waterbodies but also that of overall coastal ecosystems.

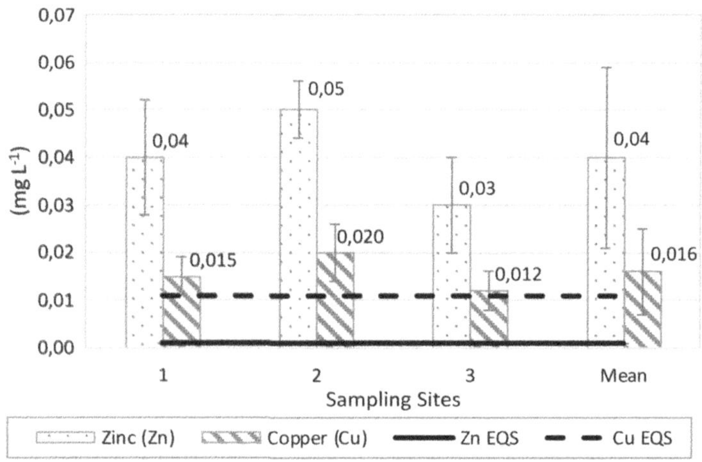

Fig. 11. Mean Copper and Zinc Concentrations (mg•L⁻¹) from Water Samples at Three Sites with ± 2 x s.d. Shown as Error Bars. EQS of Cu (0.011 mg•L⁻¹) and Zn (0.001 mg•L⁻¹) Shown as Horizontal Lines.

The reference legislation is the EU Water Framework Directive (2000/60/EC) which introduced the River Basin Management Plan (RBMP), setting out legally binding objectives, standards, and measures for all economic sectors. Each RBMP must apply to a RBD, an area made up of one or more neighbouring river basins (Fig. 12) (DEFRA, 2021).

The investigated area belongs to the Southwest RBD. The South West River Basin Management Plan (SWRBPM) considers the significant water management issues and how these could be managed, the most important are:

- Managing pollution from rural areas – Pollution is caused by the combined effects of numerous sources and its management benefits farmers, helps reduce localised flooding, reduces sedimentation of lakes and harbours, improves fisheries, and reduces harmful chemicals entering water bodies.
  - Solution: diffuse pollution at source. Nitrate Vulnerable Zones (NVZs) are designated where nitrate concentration in surface and/or ground waters are high or increasing. Farmers within NVZs must comply with mandatory action

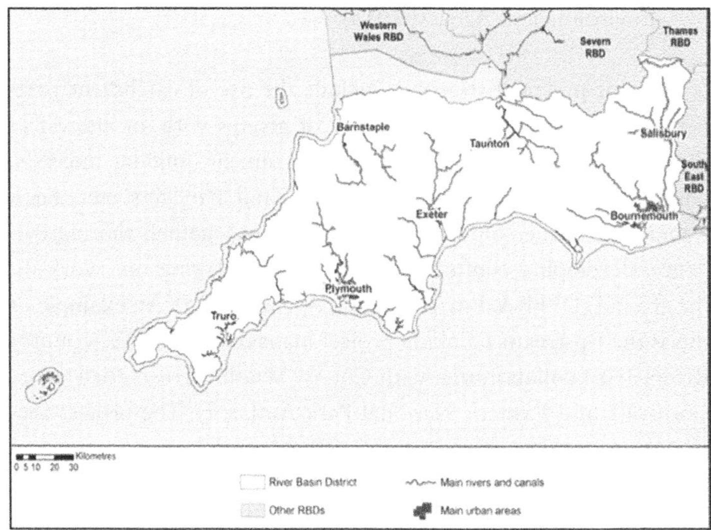

Fig. 12. Map of Southwest River Basin District (SWRBD) (DEFRA, Environmental Agency, 2011).

programme measures to reduce agriculture nitrate losses (DEFRA, Environmental Agency, 2015).

- Managing pollution from wastewater – Wastewater can contain nutrients (e.g. phosphates and nitrates), harmful chemicals (e.g. ammonium and metals), and other harmful substances (e.g. viruses and bacteria).
    - Solution: mitigate and remediate point source impacts on the receptor, reduce diffuse pollution at source, and reduce point source pathways (i.e. control entry to water environment) (DEFRA, Environmental Agency, 2015).
- Physical modification
    - Solution: removal or modification of engineering structures, vegetation management, removal of barriers to fish migration, and improved conditions of the riparian zone and wetland habitats. There is increasing evidence, that in some cases, addressing the impacts of modifications (e.g. using natural water retention measures such as wetland creation and coastal realignment) could help elevate flooding by slowing flows and making more space for water (DEFRA, Environmental Agency, 2015).

Current practical strategies include the use of catchment partnerships, encouraging a wide range of groups with an interest in the water environment (e.g. local government, angling interests, wildlife organisations, water companies, land managers, etc.). Each catchment partnership commits to working together, sharing evidence, developing common priorities, and carrying out work on the ground (DEFRA, Environmental Agency, 2015). An example of this is the Upstream Thinking project managed by the Westcountry Rivers Trust collaborating with County Wildlife Trusts for Devon, Cornwall, and Exmoor National Park Authority. The project focuses on addressing eutrophication issues in catchment waters and preventing the entry of pollutants (South West Water Limited, n.d.). The project is carried out by working closely with farmers, distributing soil testing kits and specific land and water management plans. Additionally, they are granted a pesticide amnesty to dispose of banned and out-of-date pesticides in target areas. Until now the project has been able to remove nearly a tonne of chemicals (South West Water Limited, n.d.). This plan can be further developed by establishing more fertiliser regulations, regulating both their quantity in the agricultural land and the nearby water streams.

A key issue is funding sources: the Water Services Regulation Authority (Ofwat) reviews water industry investment plans every five years (2015–2020 £3.5 billion in environmental improvements across England and Wales). Significant investment goes into point source impacts from sewage treatment works and discharges from the sewer network, while further funding targets obstruction and flow pressures (DEFRA, Environmental Agency, 2015).

Additional catchment land use and industrial management strategies could include:

- good farm slurry and waste management to ensure sufficient capacity and leakage prevention,
- the use of riparian buffer zones to prevent pollution runoff in streams,
- implementing methods of runoff diversion, e.g. barriers,
- encouraging the use of healthier NPK fertilisers (nitrogen, phosphorus, and potassium) including testing prior to use,
- the spreading of organic manures during dry periods to minimise rain wash into streams.

The use of bio-monitoring can be a simple and effective measure to evaluate the ecological status and notice changes following the different strategies on the aquatic fauna.

The removal of heavy metals from water bodies should also be evaluated, methods include current technologies for filtering, precipitation, and ion exchange; these are effective but costly. An economic alternative is a technique of using natural and inexpensive minerals with the capacity to absorb (sorption) metals (Green2Sustain, n.d.).

Improvement

The inclusion of climate projections and scenarios should improve RBMP, despite introducing uncertainties in models. Nevertheless, sustainable adaptation and mitigation strategies for the management and reduction of climate change risks must be introduced, also avoiding counterproductive measures for the water environment or for the resilience of water ecosystems (European Commission, 2009).

Adaptation to climate change will be required across Europe. Possible future impacts in one river basin might resemble the current situation in another, and useful adaptation experiences in other regions with similar characteristics might exist. It is important to exchange knowledge and experiences with other RBDs and integrate lessons learned. Identifying stakeholders, working in partnership, and ensuring communication and coordination on climate change adaptation issues are key to reaching integrated and coordinated solutions (European Commission, 2009).

Conclusion and Future Work

The investigation of the Wembury streams indicated several breaches of EQS: water nitrogen, copper and zinc, and sediment copper. Only three of the seven measured parameters complied with EQS. The highest recorded concentrations were mainly at sites one and two, likely from point source inputs from Wembury town and pollution accumulation from upstream land use such as arable and agricultural land. A comparison with river Erme to assess whether the data was common for the area showed that it also had EQS breaches but to some degree displayed an overall better ecological status. There are several management policies among the SWRBMP, Government and EU legislation, together with great participation from local partnerships, all collaboratively working towards improving the local environment.

It is recommended to regularly perform sample testing and biomonitoring of nutrients and metal concentrations in the catchment, together with strengthening measures to decrease the runoff of nutrients and other pollutants. Additionally, an in-depth investigation upstream within the river system should be performed to identify point sources of pollution and to implement the 'polluter pays principle'.

A special precaution must be taken for sediment copper, increasing monitoring to ensure values do not exceed PEL (Table 1) possibly becoming dangerous for the fauna and flora but also for humans.

It is worth noting that EQS are set to mimic pristine conditions which are hard to achieve in practice. Despite several breaches in the legal limits, Wembury displays an overall good ecological status supporting life above and below water and is therefore an appropriate model for promoting environmental stewardship.

REFLECTIONS

These case studies are not only great practice for environmental assessment and report writing but above all for developing practical skills, as students enter the field and work on collected data. This assignment showed how sensitive water quality is to anthropogenic influences, stressing the importance and relevance of water quality practice. A stronger involvement of students in this specific subject could become relevant for covering vacancies, for instance in the Environmental Agency, further providing a regular flow of data which could be publicly used, and on a longer-term view, influence public opinion and future generations on the importance of active water management and maintaining a sustainable agenda.

On a separate note, related to academic learning techniques, the University of Plymouth incorporated the Problem-Based Learning (PBL) approach in its curriculum, encouraging the application of knowledge to real-world scenarios. The author was very influenced by the PBL experience and its link with the SDG14 initiative. This learning approach can be very effective as it immerses the students in real issues, practical solutions must be developed, preparing students for a professional working environment.

A great focus is placed on the three pillars of sustainability: environmental, economic, and societal relevance. When addressing PBL issues, it is fundamental to incorporate all three pillars in a balanced manner to ensure long-term sustainability. PBL is a great tool to show the application of knowledge and to allow students to become the stakeholders and guardians of tomorrow's world.

Pollution, climate change, loss of the biological integrity and habitat of the planet, and acidification of the oceans and water bodies including river systems and estuaries undoubtedly violate the right of humankind to a healthy environment but there are no specific international standards that expressly sanction this obligation.

Water is the pinnacle of life, humans and the world we live in need water to survive and contribute to our overall wellbeing. Naturally, all SDGs concerning water are strictly interlinked, in particular from SDG14, the one closer to the author, to SDG 6 (Clean Water and Sanitation) and 3 (Good Health and Well-being). A clean and healthy water supply and ecosystem are among the strongest enablers of human health, linking with the activities and ultimate goals of an organisation such as the World Health Organization.

REFERENCES

British Geological Survey. 2020a. *Copper (Cu) in Stream Sediments.* Available at: https://www.bgs.ac.uk/download/copper-cu-in-stream-sediments/ [Accessed 3 January 2022].

British Geological Survey. 2020b. *Zinc (Zn) in Stream Sediment.* Available at: https://www.bgs.ac.uk/download/zinc-zn-in-stream-sediments/ [Accessed 3 January 2022].

Canadian Council of Ministers of the Environment (Environment). 1999a. *Canadian Sediment Quality Guidelines for the Protection of Aquatic Life – Copper*, Ottawa, Canadian Council of Ministers of the Environment [Accessed 18 November 2021].

Canadian Council of Ministers of the Environment (Environment). 1999b. *Canadian Sediment Quality Guidelines for the Protection of Aquatic Life – Zinc*, Ottawa, Canadian Council of Ministers of the Environment [Accessed 18 November 2021].

Canadian Council of Ministers of the Environment (Environment). 1999c. *Canadian Water Quality Guidelines for the Protection of*

Aquatic Life – Nutrients: Canadian Guidance Framework for the Management of Nearshore Marine Systems, Ottawa, Canadian Council of Ministers of the Environment [Accessed 18 November 2021].

Cirinà, V. 2021. *Wembury Catchment. 1:25000. Land Cover Map 2015 UK Centre for Ecology & Hydrology. Using: QGIS 3.22.1 Białowieża*, Plymouth, University of Plymouth.

DEFRA, Environmental Agency. 2011. *Water for Life and Livelihoods – River Basin Management Plan South West River Basin District*, Bristol, Environmental Agency [Accessed 20 December 2021].

DEFRA, Environmental Agency. 2015. *Part 1: South West River Basin District – River Basin Management Plan*, Bristol, Environmental Agency [Accessed 20 December 2021].

DEFRA, Environmental Agency. 2019. *2021 River Basin Management Plan*, Bristol, Environmental Agency [Accessed 5 January 2022].

DEFRA, Environmental Agency. 2021. *Erme Water Body*. Available at: https://environment.data.gov.uk/catchment-planning/WaterBody/GB108046005200 [Accessed 4 January 2022].

DEFRA, Environmental Agency. 2021. *River Basin Planning Guidance*, Bristol, Environmental Agency [Accessed 26 December 2021].

DEFRA, Environmental Agency. 2022. *Water Quality Archive – Sampling Point – River Erme at Sequers Bridge*. Available at: https://environment.data.gov.uk/water-quality/view/sampling-point/SW-70920104?_all=true [Accessed 4 January 2022].

Department of Environment and Conservation. 2010. *Contaminated Sites Management Series – Assessment levels for Soil, Sediment and Water*, Bentley, The Government of Western Australia – Department of Environment and Conservation [Accessed 13 April 2021].

Elkhatat, A. M., Soliman, M., Ismail, R., Ahmed, S., Abounahia, N., Mubashir, S., Fouladi, S., & Khraisheh, M. 2021. Recent trends of copper detection in water samples, *Bulletin of the National Research Centre*, *45*, 218.

European Commission. 2009. *Common Implementation Strategy for the Water Framework Directive (2000/60/EC) – Guidance Document*

No. 24 River Basin Management in a Changing Climate, Brussels, European Commission.

Green2Sustain. n.d. *SORPMET: Sorption of Metals by Lows Cost Natural Materials*. Available at: https://www.green2sustain.gr/sorpmet-sorption-of-metals-by-lows-cost-natural-materials/ [Accessed 26 December 2021].

Mindat.org. 2022. *Emily Mine (Wheal Emily), Wembury, South Hams, Devon, England, UK*. Available at: https://www.mindat.org/loc-61417.html [Accessed 4 January 2022].

Mousavi, S.R. 2011. Zinc in crop production and interaction with phosphorus, *Australian Journal of Basic and Applied Sciences*, 5(9), 1503–1509.

NOAA. n.d.A. Harmful Algal Blooms. Available at: https://oceanservice.noaa.gov/hazards/hab/ [Accessed 10 April 2022].

NOAA. n.d.B. *What Is Eutrophication?* Available at: https://oceanservice.noaa.gov/facts/eutrophication.html [Accessed 10 April 2022].

Rieuwerts, J. 2015. *The Elements of Environmental Pollution*, London, Routledge.

South West Water Limited. n.d. The Project. Available at: https://www.southwestwater.co.uk/environment/working-in-the-environment/upstream-thinking/the-project/ [Accessed 7 March 2022].

UK Technical Advisory Group on the Water Framework Directive (UKTAG). 2013a. *Updated Recommendations on Environmental Standards – River Basin Management (2015-21)*, London, UKTAG [Accessed 19 December 2021].

UK Technical Advisory Group on the Water Framework Directive (UKTAG). 2013b. *Updated Recommendations on Phosphorus for Rivers – River Basin Management (2015-2021)*, London, UKTAG [Accessed 19 December 2021].

3

IMPACT ASSESSMENT OF A SUBSEA TIDAL GENERATOR ON BIODIVERSITY

Matt Elliott Bell

The University of Plymouth, UK

ABSTRACT

Divers were contracted to carry out a detailed baseline survey which will form the Environmental Impact Assessment. This report presents information about the biodiversity of Cawsand Bay and the impact of installing a subsea tidal energy module. Subsequently, this addresses some of the SDG14 targets: 14.5, conserve coastal and marine areas; 14.7, increase the economic benefits from the sustainable use of marine resources to small island developing states and less developed countries; and 14.8, increase scientific knowledge, research and technology for ocean health. Contracted from November to December 2021 over a four-week period, five SCUBA divers conducted baseline transects over regular intervals of five meters at Cawsand Bay in each cardinal direction. Water and sediment samples were analysed to better understand the habitat and benthos at Cawsand Bay. Sediment samples established the biotope by identifying the benthos: sublittoral seagrass beds (SS.SMp.SSgr.Zmar). The data also revealed *Zostera marina*, commonly known as eelgrass (seagrass), is the most abundant species in the area, resulting in a high oxygen content within the water samples. In turn, this helps establish an environment capable of sustaining high levels of biodiversity for this time of year and is a more efficient support ecosystem.

Keywords: Renewable energy; subsea; baseline survey; biodiversity; sustainability; seagrass

UNIVERSITY, THE FOREFRONT OF KNOWLEDGE

Within the University of Plymouth, the author has found a purposeful and versatile learning environment. The lecturers and technicians' enthusiasm for passing on their knowledge and inspiring the next generation is palpable. The University of Plymouth, with its flotilla of boats, Remotely Operated Vehicle's (ROV's) and field equipment, has the added bonus of enabling students to collect data to form the basis of their own reports for assessments and pushing humanity's understanding of life below water through sustainable agendas. These are the main reasons for the author to choose to study at the University of Plymouth, but what the author loves most is the fusion of a scientific core with an exploratory nature.

In this chapter, the author will talk about his experience in higher education from an undergraduate perspective. Sustainable Development Goal 14 can be achieved not only in higher education but also in secondary education. The author was first introduced to the SDGs in his IGCSE and A-Level Geography curricula. It is within the school setting that the author was introduced to the United Nations' role and how we can begin to tackle current big world issues.

It is the author's opinion that there is no reason this process cannot start in the classroom: exams and assessments illustrate an understanding of the core principles but often do not fully engage the students in the creative mindset of sustainability required to tackle real-world problems. Advancement to higher education provides the opportunity to dive into the subject at a granular level.

The following case study has been taken from the 'Scientific Diving Module', OS207. One of the assessments undertaken has been reviewed and reflected upon to ascertain how the taught principles and assessments link with SDG14 in higher education. The author intends to show that lectures and assessments, set by the university, support SDG14 and prepare the scientists of tomorrow to have a real-world impact.

AN IMPACT ASSESSMENT OF A SUBSEA TIDAL GENERATOR ON LOCAL MARINE BIODIVERSITY

Cawsand Bay, Plymouth Sound Dive Survey

INTRODUCTION

Cawsand Bay is a shallow, soft and fine sediment bay found on the west side of Plymouth Sound (Fig. 1). It is sheltered by Rame Head from prevailing south-westerly winds and large tidal currents that carry across Plymouth Sound. This creates a sheltered diving environment in variable conditions. Cawsand Bay has a tidal range of 5.8 m (Fig. 2), estimated from the nearest port, Devonport, Plymouth, which slightly alters the tidal times. The precise location of the survey was N 50′19°.737 and W 4′11°.742 (Fig. 1). The bay has historical significance, and its natural proclivity to provide shelter against storms meant it was used heavily by smugglers in the 1830s. Until the completion of the Breakwater in 1844, the Royal Navy used the bay for harbouring. In more recent times, Cawsand Bay has been home to over 100 moored fishing vessels (Keble, 2006).

Plymouth Sound and Estuaries Special Area of Conservation (SAC) covers and protects areas of the Sound to allow for better sustaining of the biotope and ria systems (Marine Management Organisation [MMO], 2019). Cawsand Bay sits within this SAC. Although a rare angiosperm, *Zostera marina* is situated along the coasts of Great Britain and is one of the most predominant species to grow within the soft sediment at Cawsand Bay, contributing towards oxygen-rich waters and an environment capable of sustaining high biodiversity within the Plymouth Sound.

To examine the effects of installing a subtidal energy generating module, a group of five divers surveyed the same shot line in Cawsand Bay, situated to the North-East of Pier Cellars (torpedo

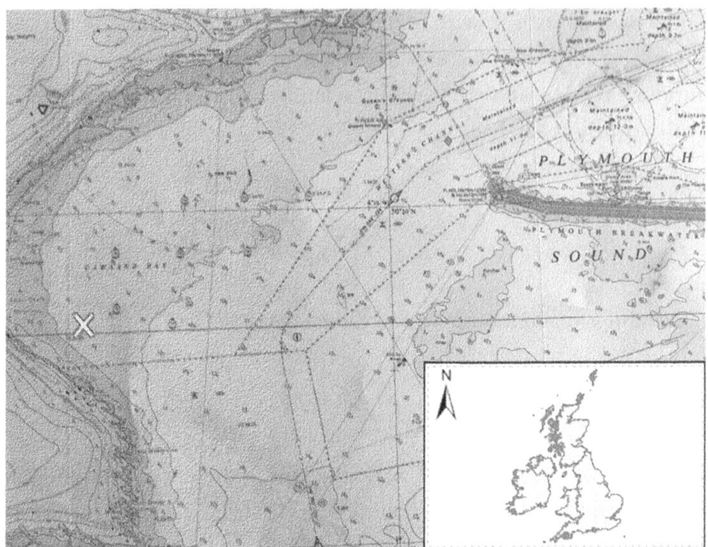

Fig. 1. Map Depicting Baseline Survey Site Marked with an 'X'. Coordinates: N 50'19°.737 W 4'11°.742 (Admiralty Chart – 30 Plymouth Sound and Approaches, UK Hydrographic Office) (R Core Development Team, 2020, R version 4.0.3 (2020-10-10)).

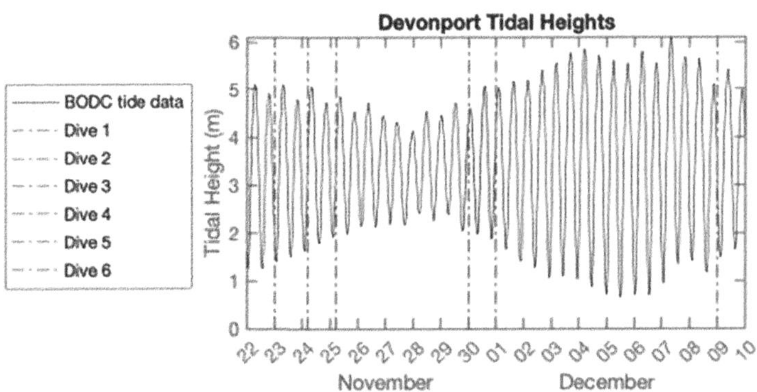

Fig. 2. Tidal Chart for the Survey Period. Dives Took Place Over Varying Tidal States (National Oceanography Centre, 2021).
Note: **As tide data is unprocessed, it may lead to minor irregularities, i.e. the storm surge observed on the 8th of December.**

launch site) (Fig. 1), over a series of four weeks. They collected data on the biodiversity, sediment type and water samples, all key parameters contributing to understanding life below water

(SDG14). To collect data, transects were taken in northern, eastern, southern and western directions from the shot. All data were collected at the same location over the duration of the project for consistency and repeatability. These data provide a foundation for understanding the biodiversity present in Cawsand Bay; the water and sediment samples taken on dives provide data on the benthos and type of habitat that surrounded these organisms.

Sediment samples provide quantitative data on the grain size, sediment type and the foraminifera of Cawsand Bay. To further understand benthic communities within the sediment, the abundance of different foraminifera was identified down to family and where possible to species level. In contrast, water samples were taken to obtain information on water quality, in particular salinity. Oxygen is a limiting factor governing the level of biodiversity present in the area. In saline water, oxygen solubility is 20% lower than in freshwater due to the presence of sodium and chloride ions.

METHODS

Prior to data collection, the methodology was tested on a preliminary dive. Incidences of species misidentification were reduced by the team's construction of a dive slate containing identifiable characteristics of species. To achieve accuracy and repeatability, this was used over the four-week data collection period. However, periods of unworkable weather could affect the data collection process as did poor visibility. The latter could negatively impact the representation of the percentage covered displayed in the photo-quadrats of *Zostera marina* and other benthic species. In such instances, data was solely extracted from the second diver's observation. As the shot dive method was redeployed for each dive, the skipper reidentified the location using meridian navigation (latitude and longitude).

Preliminary Dive

The initial reconnaissance and data collection for all four weeks was carried out as shown in Fig. 3. The first dive was conducted to familiarise divers with the study area and allowed for the

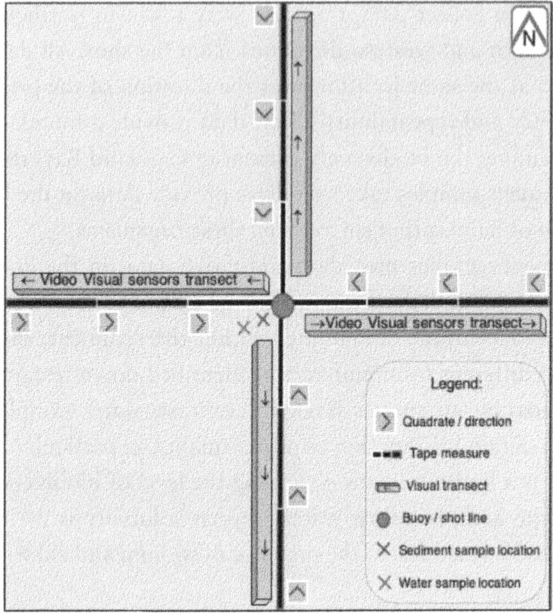

Fig. 3. Plan View of Survey.

reconnaissance of potential risks below and above water on the dive site. Any risks found were added to the risk assessment form. All dives were carried out using a Canon PowerShot G9 X Mark II camera. Recording data in this way facilitated group discussion and more time above water to do species ID. A video transect was recorded 15 m in each cardinal direction. Upon their return to the shot, divers took 0.5 m^2 photo-quadrats at five meter intervals. This process was repeated each dive for the four-week process.

Sediment Sample

Sediment samples were collected to identify the composition of the sediment particulate size. Prior literature observes that physical characteristics of sediment such as particle size influences the benthic community; a larger particle size increases species abundance (Huotari, 2015; Rhoads and Boyer, 1982). During each dive, divers used a sediment pot to extract sediment at the shot to identify the sample's exact GPS location (Fig. 3).

Sediment Analysis Method

The sediment was prepared by mixing in rose bengal to stain the living tissue. This was left in situ for 24 h. Samples were then emptied into a 63 µm sieve, as shown in Fig. 4. Water was used to remove particulate matter and rose bengal from the sediment and push the smaller sediment through the sieve. Once filtered, the sediment was tipped onto two metal trays and left to dry in the oven (Fig. 4). Sediment samples were put through testing sieves to separate particle sizes (Fig. 4). The dry sediment was then weighed at each sieve size to obtain the benthic composition of the sediment grain size (Fig. 5). The sediment sample from 125 µm was analysed under a light microscope (Fig. 4).

Water Samples

In line with Boyle's Law whereby pressure and gas have an inverse relationship, sample bottles were filled with seawater prior to the dive. Upon reaching the site, divers descended the shot; one diver took the sediment sample while the other, the water sample (Fig. 3). The bottle was inverted over the diver's exhaust valve, filling it with air and removing the seawater. This step was repeated three times to remove any contamination from surface seawater. Water samples were taken on every dive at the bottom of the shot to keep data collection and analysis consistent. After each dive,

Fig. 4. 63 µm Wet Sediment Sieve (Left). Dry Testing Sieve 500 – 63 µm (Middle). Oven Used to Dry Sediment (Mid-Top Right). Sediment Species Collected (Right).

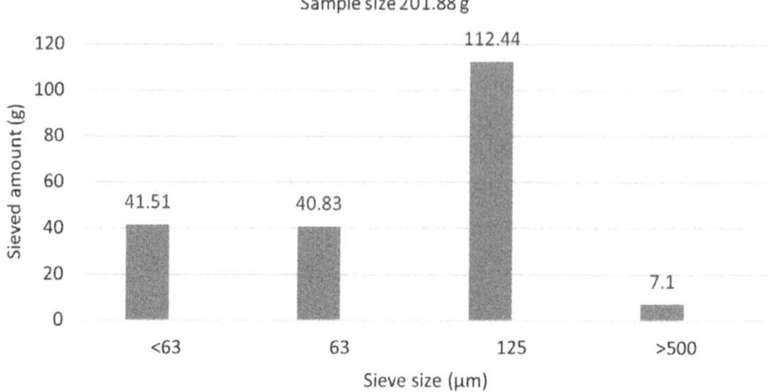

Fig. 5. Sieved Sediment Sample Distribution. Biotope Identification of the Area Classified as (SS.SMp.SSgr.Zmar).

a multiparameter water sensor (YSI model 85 water measurement probe) was used to analyse the properties of the water found at the dive site in Cawsand Bay. The probe gives a digital display of the water sample's salinity, temperature and dissolved oxygen percentage (CTDO). By comparing samples, the influences of weather variables are ruled out.

JNCC, Visual/Video/Photo Census

The first diver recorded a video in each direction on the way out, took photographs of quadrats on the way in, and repeated in each direction. This was documented using a Canon PowerShot G9 X Mark II camera, making it possible to review data and complete the Joint Nature Conservation Committee (JNCC) form. Although this reduced anomalies, occasionally poor visibility and unworkable weather affected the data collection process. The dive buddy carried out a visual census in accordance with the JNCC form to record the percentage coverage of *Zostera marina*, algae and species ID on the diving slate. In times of poor visibility, the second diver was relied upon. However, it was not possible to cross-check this data with the team.

Quadrat Samples

A Canon PowerShot G9 X Mark II was used to photograph 0.5 m^2 quadrats every five meters, using the 15 m transect line as a reference point to identify species diversity. Images were taken directly over the quadrat to prevent obstruction of view. After one cardinal direction was complete, divers would swim onto the next cardinal direction and repeat the sampling procedure (Fig. 3). This was to ensure a systematic sampling procedure that would prevent repeat surveying to reduce quadrat distortion for analysis and provide a profile of species along the entire sample site. The quadrats were analysed by phyla, and where possible, species level was identified within each photo-quadrat. Original data were also compared to sample database quadrats taken in 2020. These data are of the same area but sampled using a jackstay survey method. Therefore, it is not a fully comparative data set and is only referred to as a baseline survey for biodiversity indexes.

Statistics and Software Used

The hypothesis: seagrass coverage is significantly different by direction.

The null hypothesis: seagrass coverage is not significantly different by direction.

Statistical analysis of data was conducted in Microsoft Excel 2021 and MatLab version 9.10.0.1649659 (R2021a). A one-way ANOVA was used to test if seagrass or algae percentage coverage varied significantly by direction within the survey area.

The marine mammal animal index was calculated in MatLab to determine the biodiversity in infaunal and benthic communities.

The Shannon–Weiner index (H) describes the disorder between species. The higher the disorder, the higher the biodiversity.

$$H = -\sum_{i=1}^{s} p_i \ln(p_i)$$

where p_i = decimal proportion of species.

Evenness measure (E) is calculated using the Shannon evenness index. It is used to compare the number of species and their abundance.

$$E = \frac{H}{H_{max}}$$

where H = Shannon–Wiener index, H_{max} = ln species number.

The Simpson index (D) measures the dominant species in a multispecies habitat. It shows the probability of two randomly selected individuals from the sample being from the same species.

$$D = 1 - \left(\frac{\Sigma n(n-1)}{N(N-1)}\right)$$

where N = total number of all organisms and n = number of organisms within a certain species.

RESULTS

Marine Plant and Algae Diversity

Results from the one-way ANOVA:

Seagrass directional P value = 0.938

Macroalgae directional P value = 0.461

Within this studies survey area there is no significant difference in the percentage coverage of seagrass or macroalgae by direction. There is also even coverage of seagrass and macroalgae over the area surveyed.

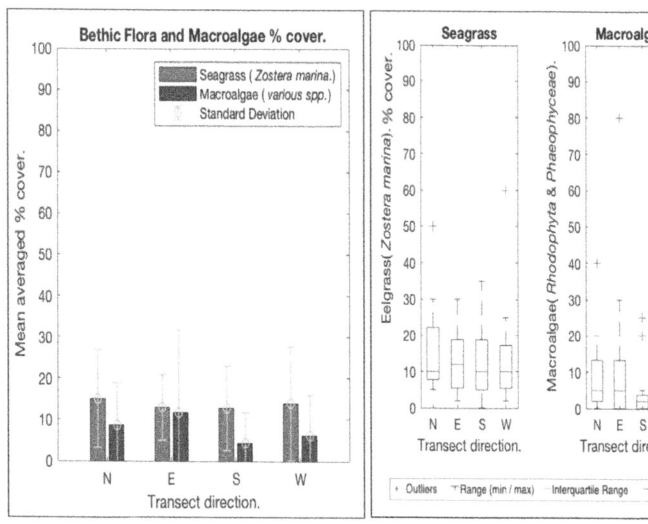

Fig. 6. Percentage Cover of Seagrass and Macroalgae Species. Error Bars Show 95% Standard Deviation (Left). Box Plot Shows a Five Number Statistical Summary of the Data to Provide a Measure of Reliability for Quadrat Data Samples (Right).

Marine Animal Diversity

Table 1. Count from Sediment and Visual Surveys.

Common name	Scientific name	Number of species
Annelid worm	*Annelid worm*	1
Shore crab	*Carcinus manas*	9
Conger eel	*Conger conger*	1
Spider crab	*Hyas araneus*	1
Hermit crab	*Pagurus bernhardus*	18
Sand goby	*Pomatoschistus minutus*	7
Fan worm	*Sabella spallanzanii*	29
Foraminifera (Hyaline)	*Ammonia beccari*	6
Molluscs	*Bivalves* (SP)	7
Crustaceans	*Crustaceans shed* (sp)	1
Foraminifera (Hyaline)	*Elphidium williamsoni*	47

(Continued)

Table 1. (*Continued*)

Common name	Scientific name	Number of species
Marine snail	*Gastropod* (sp)	1
Foraminifera (Porcelaneous)	*Massilina secans*	34
Ostracod (sp)	*Ostracod* (sp)	1
Blue-rayed limpet	*Patella pellucida*	1
Limpet	*Patellidae* (sp)	2

Notes: Black text (Annelid worm to Fan worm) species indicate visible species; gray text (Foraminifera to Limpet) indicates species identified under the microscope. Sizes vary upward of 125 µm, this being the common sediment size (Fig. 5).

Table 2. Biodiversity Indices from Both Sediment Dwelling Communities and Benthic Habitats.

Biodiversity indexes from visual survey	Biodiversity indices for in-fauna (sediment dwelling species)	Biodiversity indices for epi-fauna (benthic dwelling species)
Richness	9	7
Total number of individuals sampled	105	75
Average population size	11.7	10.7
Shannon-Weiner diversity index (*H*)	1.441	1.49
Evenness (*E*)	0.656	0.768
Simpsons Index (*D*)	0.317	0.265

Notes: Data from 2020 base line survey at the site was compared to the present study's data.
The Shannon–Weiner index: 1.14.
Evenness measure: 0.46.
The Simpson index: 0.54, a probability of 54% of individuals are from the same species.

Water Sample Results

Table 3. Water Sample and Statistical Analysis Results.

2020 Water Sample Data

Data set	Depth	Temp (°C)	Sal (PSU)	pH	O_2 %	O_2 (mg.L^{-1})
09/12/2020	0m	11.10	30.10	-	-	16.91
09/12/2020	2m	11.60	31.20	-	-	17.32
09/12/2020	5m	11.50	30.70	-	-	5.10
09/12/2020	8m	11.70	32.30	-	-	16.70
09/12/2020	12m	12.10	32.10	-	-	17.50

Statistics 2020

Variable	Mean	SE Mean	Sthev	Variance	Minimum	Median	Maximum	Range
Temp (°C)	11.6	0.161	0.361	0.13	11.1	11.6	12.1	1
Salinity (PSU)	31.28	0.415	0.928	0.862	30.1	31.2	32.3	2.2
Oxygen (mg.l^{-1})	14.71	2.41	5.38	28.94	5.1	16.91	17.5	12.4

(Continued)

Table 3. *(Continued)*

2021 Water Sample Data

Data set	Depth	Temp (°C)	Sal (PSU)	pH	O_2%	O_2 (mg.L^{-1})
09/12/2021	10m	13.50	34.16	7.94	103.10	8.70
01/12/2021	8m	14.30	34.09	7.77	100.60	8.35
30/11/2021	9m	12.90	33.71	7.62	95.90	8.17
25/11/2021	9m	12.40	34.45	7.95	96.30	8.30
24/11/2021	9m	11.90	33.19	7.94	98.40	8.63
23/11/2021	9m	14.20	34.01	7.79	98.50	8.19

Statistics 2021

Variable	Mean	SE Mean	StDev	Variance	Minimum	Median	Maximum	Range
Temp (°C)	13.2	0.397	0.972	0.944	11.9	13.2	14.3	2.4
Sal (PSU)	33.935	0.178	0.436	0.19	33.19	34.05	34.45	1.26
pH	7.835	0.0541	0.1325	0.0176	7.62	7.865	7.95	0.33
Oxygen %	98.8	1.11	2.71	7.33	95.9	98.45	103.1	7.2
Oxygen (mg.l^{-1})	8.39	0.0916	0.2244	0.0504	8.17	8.325	8.7	0.53

DISCUSSION

Sediment Sampling

The dry sediment analysis further clarifies the suitability of the benthos to place a subsea tidal energy module. Samples collected over the four-week process clearly identified the largest proportion of sediment size (125 µm), classified as fine sand on the Wentworth scale (Fig. 5) (Wentworth, 1922). This size is ideal for benthic organisms with *Elphidium willamsoni* (47) and *Massillina secans* (34) foraminifera making up most of the collected sediment species (Table 1). The previous year's visual census noted a high Annelida count with multiple burrows in the soft sand benthos. However, several subbenthic organisms and the identification of species were not possible.

Water Sampling

The average salinity of the world's oceans is 35 partial salinity unit (PSU), which is 1.11 PSU higher than this study's average samples (33.89) (Table 3). This slight difference in salinity is most likely due to Cawsand Bay's proximity to the Tamar and Plym Estuaries, which output freshwater into the Plymouth Sound basin with fluvial flow rates of approximately 250 $m^3 s^{-1}$ (Uncles et al., 2007). Water salinity is determined by dissolved salt inputted into the system by weathering of exposed rock, mainly made up of Potassium (K^+), Chloride (Cl^-) and Sodium (Na^+) (Touchette, 2007). The location of Cawsand Bay, although protected from poor weather, can face changes in salinity, so the density and solubility of the water affect the organisms located within the sampling zone. Consequently, organisms need to be able to adapt to slight changes in accordance with salinity fluctuations (Touchette, 2007). Oxygen concentration is another key parameter that affects organisms found within the study area. The high abundance of *Zostera marina*, a photosynthesizing plant, and other macroalgae (Fig. 6 and Table 3) affects the percentage of oxygen present in the water, with a mean of 98.8% (8.39 mg. l^{-1}). The temperature of the water will affect the oxygen percentage as it becomes more soluble as evident in Table 3. The lower the temperature, the higher the oxygen solubility (Fondriest Environmental, Inc, 2013). Another factor is atmospheric pressure, which can have a profound influence on dissolved oxygen concentrations.

JNCC, Visual/Video/Photo Census

The sandy fine sediment biotope of Cawsand Bay allows for easy mobility within the benthic zone. With no sublittoral/intertidal rocky area, there is no suitable habitat to support the growth of kelp. The sediment type and species (Fig. 5 and Table 1) help define the biotope type – sublittoral seagrass beds (SS.SMp.SSgr.Zmar). The biotope of Cawsand Bay is well suited for *Zostera marina*, *Saccharina latissima*, *Pagurus bernhardus*, *Elphidium williamsoni*, *Massillina secans*, and *Sabella pavonina* as these species are proven to be most abundant within the sample. *Zostera marina's* and *Saccharina latissima* spatial distribution is abundant due to a more specialised approach to this habitat. The observation of the visual and video census indicated that *Sabella pavonina* is predominantly located near *Saccharina latissima*. In contrast, *Pagurus bernhardus* is less abundant and specialised with a weaker habitat preference for biotopes. Crustaceans such as *Pagurus bernhardus* are more generalists and require less from the habitat, allowing the exploitation of different biotopes (Freeman and Rogers, 2003). The presence of worms would suggest that the mean sediment diameter was around 125 µm; as less than this would prevent the formation of small worm burrows (Sun et al., 2019). Based on data collected at one point in the year, this report cannot provide a representative insight into the area's seasonal biodiversity. As the number and presence of species will vary in abundance over the year in accordance with seasonal breeding or spawning, future surveys of Cawsand Bay should emphasise seasonal data collection. Identifying species migratory patterns and cycles could help prevent further damage to the globally rare *Zostera marina* beds.

Biodiversity

The indexes presented in Table 2 show the overall biodiversity in 2021 was slightly higher than in 2020, but both data sets present low biodiversity overall. This is to be expected due to the large abundant diversity present in soft sediment environments found infauna. Shannon–Weiner and evenness indexes were higher in 2021 than the baseline, meaning that the data showed a higher amount of disorder (biodiversity) between species but still relatively high evenness. The Simpson index, however, is higher in the baseline data. The Simpson index is a similar measure of diversity to Shannon–Weiner but needs a larger data set and takes less account of rarer species. Although measures were comparative, it is clear from quadrat pictures that the baseline area slightly varies from the 2021 data set, so its use as a baseline was done so with caution. The measures are most likely to vary dramatically throughout the year, coinciding with various species' seasonal habits.

Noise impact was not addressed within this initial baseline survey. However, with increasing studies identifying that marine organisms are affected by anthropogenic sounds and light, it is vital that future investigations consider anthropogenic noise pollution when conducting marine surveys. Methods to dampen anthropogenic sounds from offshore wind farms and other marine constructions is an area of research requiring much investigation. Another area requiring novel and future cross-collaboration interdisciplinary research is the impact and effects of electromagnetic fields produced by offshore energy production. This was recently highlighted by Harsanyi et al. (2022) study towards the negative effects this is having on marine life, particularly benthic species, such as the European lobster (*Homarus gammarus*) and common crab (*Cancer pagurus*).

Quadrat Samples

The identified biotope comprises soft fine sand (Fig. 5) (SS.SMp.SSgr.Zmar). Although Cawsand is in a sheltered area, it is still subject to strong tidal scouring, which causes high amounts of hydrodynamic stress because of the low stability of the benthic substrate. Recently, the Ocean Conservation Trust with the National Marine Aquarium deployed five new eco moorings in Cawsand Bay. Therefore, using a jackstay at the baseline survey would have increased precision monitoring and help determine the impact of LIFE ReMEDIES Project on *Zostera marina* over time. Future studies could analyse the impact of the anthropisation of the surrounding land and evaluate the effect this has on Plymouth Sound's Marine Protected Area (MPA) at Cawsand Bay. The presence of plastic pollution and species present as recorded during the baseline survey suggests future studies could also assess the effectiveness of clean-up schemes. The present study did not happen upon any crustacean collection pots. Although Cawsand Bay is frequently used for potting and recreational fishing, bait in pots and catch drawing scavengers to the area will influence the organisms present. As shown in Fig. 2, Cawsand Bay is sheltered with limited access to tidal movement, and as such, any gain from tidal energy is likely to be overshadowed by possible ecological harm detrimental to life below water.

CONCLUSION OF COURSEWORK

The data provided to the contracting organisation will accurately estimate the local biodiversity and health of the available resources. Great emphasis was placed on maintaining a consistent methodology to generate reliable results so the company may recommend the placement of a subsea tidal energy module to its client. As no core samples of the sediment were used, this study is unlikely to have demonstrated the true benthic biodiversity and descending sediment type. Consequently, the present study leaves further avenues suitable for investigation. Not only does this include core sampling, but also the implementation of other survey and data collection methods throughout the seasons where the abundance of *Zostera marina* is likely to be higher. *Zostera marina* is one of the keystone species in Cawsand Bay, and its recorded sparse abundance is due to seasonal variability and the presence of mooring sites within the area, scraping benthic flora. However, various projects such as

LIFE ReMEDIES aiming to increase *Zostera marina* abundance would be negatively affected by the situation of a tidal module in the area.

Adverse weather and methodology inconsistencies have been taken into consideration to mitigate potential problems during data collection and ensure its reliability. Representative figures have been generated displaying area species abundance and percentage cover along with sediment analysis. The Simpson's and Shannon–Weiner biodiversity index was conducted to form both the epi-fauna of benthic dwelling species from visual surveys and the in-fauna from sediment-dwelling species. Both results indicated low species diversity. Overall, the data provides the necessary baseline information to conduct the Environmental Impact Assessment necessary to inform the contractor's client. In essence, knowing how technology such as aquatic energy hubs impact the marine environment is of paramount importance in working towards SDG14's sustainability of resources (14.8) and resident life below water (14.5) targets (United Nations Department of Economic and Social Affairs Sustainable Development, 2023).

AN UNDERGRADUATE STUDENT REFLECTION ON SDG14

Universities are positioned at the forefront of knowledge and allow students to back up their theoretical skills with a practical base. The University of Plymouth is one such example recognised by the Times Higher Education as the number one university in the world in its efforts to obtain SDG14 for marine impact in 2021. Learning about the SDGs, particularly life below water (SDG14), and having an opportunity to put the student's knowledge to the test has provided an invaluable experience in tackling potential or current real-world situations within a safe learning environment; the University of Plymouth's Problem Based Learning (PBL) scheme is central to this. The SDGs provide the higher education platform and, to some extent, the tertiary education system with the opportunity to develop upon the three pillars of sustainability: environmental, economic and social relevance.

Being in higher education allows the technological transfer of informed knowledge. Universities provide the grounds for students to learn and refine skills through tasks, assessment and exams. As humanity battles to understand and maintain life below water within a world hungry for blue food and energy, it is more important to implement SDGs into the curriculum. Students, the scientists of tomorrow, will give a fresh perspective on present and future dilemmas. Examples of this can be seen through projects conducted by students helping to shed light on future areas of investigation and possible sustainable solutions. The author has chosen this model coursework to demonstrate how students are taught to conduct good science underwater and areas of further research. This illustrates through good scientific enquiry how student researchers of today can contribute to the SDGs.

There are many other avenues you can undertake to become a more environmentally sustainable person and help scientists today. Citizen science is an important process of data gathering, informing and decision-making. Everyone can become a citizen scientist! If you live in the EU, help scientists by using the 'Marine litter watch tool'; through this, you will have a say in future policymaking and the 'eu-citizen.science' webpage. You can learn and help in the UK by joining Surfers Against Sewage and Ghost Fishing UK. Locally you can help the environment by conducting a beach litter pick and joining a 'SeaSurch' programme to enhance your taxonomic and other important scientific skills. Equally, Sail Britain creates scientific and creative multimedia opportunities to conduct citizen science through the educational exploration of life below water while learning to sail. In Spain, citizen science is encouraged with the MARNOBA International Convention, the International Coastal Cleanup-Spain, Ambiente Europe, Ocean initiatives, and other volunteer structures. In Portugal, the NGO Brigada do Mar helps by cleaning the coastline, developing and implementing events aimed at protecting biodiversity and conducting awareness campaigns about the threats of marine litter. In France, the non-profit organisation 'Wings of the Ocean' and in Greece, the NGO All For Blue Organization. There are many citizen science projects to get involved with, such as work experience, placement opportunities or weekend events. All of which help your understanding towards 'life below water' while pursuing your chosen career.

By implementing SDG14, a more sustainable and equitable ocean economy is achievable and underpinned by a fundamental understanding of proper marine management. Water is a fundamental ingredient in sustaining life as we know it. With a continually increasing understanding of life below water comes a greater understanding of the corresponding SDGs and global health. Students have a right to become stakeholders and help steward environmental resources. After all, life below water accounts for 71% of the world's sustainable development potential. This author believes that today's students are embracing the SDGs and are on a pathway to future global environmental sustainability.

ACKNOWLEDGEMENTS

I would like to thank Luke Pomeroy and Anna Farr, who have been instrumental in proofreading this chapter. I would also like to thank my family and Jane Lamont, a family friend, for their continuous encouragement from a young age. Finally thank you to the group two from the OS207 module and all staff at the University of Plymouth Marine Station who made this project possible.

REFERENCES

Fondriest Environmental, Inc. 2013. *Dissolved Oxygen. Fundamentals of Environmental Measurements*. Available at: https://www.fondriest.com/environmentalmeasurements/parameters/water-quality/dissolved-oxygen/ [Accessed 28 December 2021].

Freeman, S. and Rogers, S. 2003. A new analytical approach to the characterization of macroepibenthic habitats: linking species to the environment, *Estuarine, Coastal and Shelf Science*, 56(3–4), 749–764. doi: 10.1016/S0272-7714(02)00297-4

Harsanyi, P., Scott, K., Easton, B.A., de la Cruz Ortiz, G., Chapman, E.C., Piper, A.J., Rochas, C.M. and Lyndon, A.R. 2022. The effects of anthropogenic electromagnetic fields (EMF) on the early development of two commercially important crustaceans, European Lobster, *Homarus gammarus* (L.) and edible crab, *Cancer pagurus* (L.), *Journal of Marine Science and Engineering*, 10(5), 564.

Huotari, E. 2015. *Effects of Particle Size and Particle Heterogeneity on Benthic Functional Guilds in Elkhorn Slough, CA*. Master's Theses, San Jose State University. doi: 10.31979/etd.k6vh-9y5e

Keble, E. 2006. *King's Cutters and Smugglers 1700–1855*, Edinburgh, Ballantyne, Hanson & Co.

Marine Management Organisation (MMO). 2019. *Plymouth Sound and Estuaries SAC (Including Tamar Estuaries Complex SPA)*. Available at: https://assets.publishing.service.gov.uk/government/uploads/system/uploads/attachment_data/file/ 844567/Plymouth_Sound_and_Estuaries_SAC_and_Tamar_Complex_SPA_Factsheet.pdf [Accessed 28 December 2021].

Mathworks Inc. MatLab. 2021. Version 9.10.0.1649659 (R2021a) [computer program].

National Oceanography Centre. 2021. *British Oceanographic Data Centre*. Available at: https://www.bodc.ac.uk/data/hosted_data_systems/sea_level/uk_tide_gauge_network/ [Accessed 3 January 2022].

R Core Development Team. 2020. R version 4.0.3 (2020-10-10). RStudio. [computer program].

Rhoads, D.C. and Boyer, L.F. 1982. The effects of marine benthos on physical properties of sediments. In *Animal-Sediment Relations. Topics in Geobiology*, Eds P.L. McCall and M.J.S. Tevesz, pp. 3–52, Vol. 100, Boston, MA, Springer. doi: 10.1007/978-1-4757-1317-6_1

Sun, T., Li, X., An, D., Yu, H., Ma, Z. and Liu, F. 2019. The effect of substrate grain size on burrowing ability and distribution characteristics of *Perinereis aibuhitensis*, *Acta Oceanologica Sinica*, 38(12), 52–58. doi: 10.1007/s13131-019-1348-z

The Joint Nature Conservation Committee. 1995. *Marine Nature Conservation Review*. Available at: https://mhc.jncc.gov.uk/media/1037/survform.pdf [Accessed 18 December 2021].

Touchette, B.W. 2007. Seagrass-salinity interactions: physiological mechanisms used by submersed marine angiosperms for a life at sea, *Journal of Experimental Marine Biology and Ecology*, 350(1–2), 194–215.

Uncles, R.J., Stephens, J.A. and Harris, C. 2007. *Final Report on the Sediments and Hydrography of the Devonshire Avon Estuary*, Internal Plymouth Marine Laboratory report, Unpublished.

United Nations Department of Economic and Social Affairs Sustainable Development. 2023. *United Nations Department of Economic and Social Affairs Sustainable Development. Conserve and Sustainability Use the Oceans, Seas and Marine Resources for Sustainable Development. Targets and Indicators*, Department of Economic and Social Affairs. Available at: https://sdgs.un.org/goals/goal14 [Accessed 05 January 2023].

Wentworth, C.K. 1922. A scale of grade and class terms for clastic sediments, *The Journal of Geology*, 30(5), 377–392. doi: 10.1086/622910

4

SUSTAINABLE COASTAL DEFENSE CONCORDS WITH LIFE BELOW WATER

Martin J. Baptist

Wageningen University & Research, Wageningen, the Netherlands

ABSTRACT

This chapter examines the Netherlands' challenges in safeguarding its low-lying coastline against rising sea levels and the consequences of coastal defense strategies on marine life, particularly in relation to SDG14. Sea-level rise necessitates increased soft coastal defense strategies, affecting seafloor areas and marine biodiversity through sand extraction and sand nourishments. The use of hard structures for coastal defense contributes to the loss of natural coastal habitats, raising biodiversity concerns. The chapter explores the potential benefits of artificial hard surfaces as marine habitats, emphasising the need for careful design to prevent ecological problems caused by invasive species. Strategies for enhancing biodiversity on human-made hard substrate structures, including material variations, hole drilling, and adaptations, are discussed. The ecological impact of marine sand extraction is examined, detailing its effects on benthic fauna, sediment characteristics, primary production, and fish and shrimp populations. Solutions proposed include improved design for mining areas, ecosystem-based rules for extraction sites, and ecologically enriched extraction areas. The ecosystem effects of marine sand nourishments are also analysed, considering the impact on habitat suitability for various species. The chemical effects of anaerobic sediment and recovery challenges are addressed. Mitigation measures, such as strategic nourishment location and timing, adherence to local morphology, and technical solutions, are suggested. The chapter underscores the importance of education in Nature-based Solutions and announces the launch of a new BSc programme in Marine Sciences at Wageningen University & Research, integrating social and ecological knowledge to address challenges in seas, oceans, and coastal regions and support SDG14 goals.

Keywords: Coastal defense; biodiversity; hard substrates; sand extraction; sand nourishment; nature-based solutions; building with nature

INTRODUCTION

The Netherlands is a low-lying country with a high flood protection level. Most of the Dutch coast is protected by beaches and dunes that are maintained by sand nourishments at a current rate of 10–12 million m^3/yr (Lodder et al., 2019). As sea-level rise (SLR) accelerates, increasing volumes of sand will be needed for coastal defense. Anticipating an SLR of 80 cm per century, Rijkswaterstaat, responsible for water management in the Netherlands, estimates that a nourishment rate of 25–35 million m^3/yr is needed for coastal defense. However, recent advice for the Dutch Delta Commission by Deltares (Haasnoot et al., 2018) estimated that even 240 million m^3/yr might be needed by the end of this century. The hope is that the Paris Agreement will prevent this worst-case scenario from happening, but a two- to threefold increase in current sand nourishment rates is almost certain. This climate-induced increase of sand volumes for coastal defense will lead to an increase in impacted seafloor area and will directly affect life below water, i.e. habitats, biodiversity, and fishing grounds both at the extraction sites as well as at the nourishment sites.

Next to the soft-defense strategy of nourishing sand to beaches and dunes, large parts of the Netherlands are protected by hard structures in the form of breakwaters, groins, dikes, and dams. These structures create robust and highly functional long-lasting coastal defences. However, the replacement of natural coastal habitats with artificial hard substrates has come under scrutiny when considering life below water and the global issue of biodiversity decline.

To increase knowledge on ways to reduce the negative effects of both hard and soft coastal defense on marine life below water, Dutch engineers and ecologists have been working together to develop improved nature-inclusive solutions. A consortium of private parties, government organisations, research institutes, universities, and NGOs formed the Dutch foundation called EcoShape. They carry out the 'Building with Nature' innovation programme. The programme aims to test and develop a new design philosophy in hydraulic engineering that utilises the forces of nature thereby strengthening nature, economy and society (de Vriend and van Koningsveld, 2012). Building with Nature can be described as the application of Nature-based Solutions in hydraulic engineering.

Wageningen University & Research (WUR) has been one of the founding fathers of the EcoShape consortium and has been involved since 2007. It is contributing to fundamental and applied studies into the application of Building with Nature (BwN) principles. The typical role of WUR in the consortium is to bring in ecological and societal expertise, such as coupling BwN to marine aquaculture and marine governance. The Building with Nature programme has supported many PhD, MSc, and BSc students that have been active in unique large-scale field pilots. The attention to sustainability and biodiversity in the programme resulted in a new generation of young engineers with interdisciplinary skills and a dedication to developing a sustainable world.

WAYS TO PROMOTE SUSTAINABLE COASTAL DEFENSE

Constructing and maintaining coastal defences, either by hard or soft strategies, will inevitably have effects on life below water. The enhancement of underwater living conditions for biota by altering coastal defense methods and structures can make them more sustainable, which is beneficial for SDG14.

Effects of Hard Coastal Defense on Life Below Water

The rocky shore environment is known to be a highly diverse ecosystem that hosts a wide variety of functional groups of biota. The hard structures provide a surface for attachment and establishment of epibiotic species, while rock pools and crevices serve as a protected niche and refuge place (Mendonça et al., 2018). The biological community of the rocky shore is made up of macroalgae (seaweeds), grazing herbivores, and sessile fauna (permanently attached organisms). This provides food for other marine life, such as sea birds, mammals, and fish (Ellis et al., 2007).

The construction of artificial hard surfaces on a coast, therefore, presents an interesting opportunity for increasing its value as a habitat (Chapman and Underwood, 2011), Fig. 1. It is, however, important that the structures are designed and implemented in ways that provide long-lasting niches and attract desirable species. Artificial structures that do not promote natural biodiversity assemblages can potentially become hosts for invasive species that can lead to the proliferation of the invader causing cascading ecological problems.

In the case of the Netherlands, its coastal areas are exclusively made up of sandy or muddy shores. Hard coastal engineering structures form a new habitat. Generally, hard surfaces will always attract marine species; however, it is essential to look deeper into the type of species assemblages that are present and how this affects local biota.

Towards More Sustainable Hard Coastal Defense

Artificial coastal defense structures can be developed or altered in various ways to enhance local biodiversity. One of the early studies in this field was a design study by Baptist et al. (2007). They suggested varying in material type and three-dimensional shapes to increase biodiversity. Research has increased over the last decade,

Fig. 1. A Platform at Low Intertidal Height as Part of a Sea Defense Offers a High Suitability for Blue Mussel Growth. Photo by Martin Baptist.

and there are now several ways of increasing biodiversity on human-built hard substrate structures.

In an extensive review, Strain et al. (2018) assessed 109 studies that target species/biodiversity enhancement on man-made coastal structures. Their findings indicate that the largest influence on sessile organisms such as barnacles and seaweeds came from adaptations like crevices and pits. They further state that the size of the target organism should be matched with the size of the intervention. Drilling small holes into a substrate influences the abundance of smaller organisms (such as limpets), whilst larger holes and crevices influence mobile species such as fish and lobsters.

Additionally, there are both chemical and physical adaptations that can be applied. Ido and Shimrit (2015) conducted a study comparing modified concrete with standard coastal structures. During the casting process, the modified blocks were equipped with crevices and holes by applying an elastic liner. Additionally, an altered cement mix resulted in a lower pH. As a result, the modified blocks were found to harbour larger biodiversity with a lower ratio of non-native species.

Effects of Marine Sand Extraction on Life Below Water

Marine sand extraction leads to removal and mortality of benthic fauna at the extraction site, and to smothering and burial of benthos in the vicinity. The morphology of the seabed changes, which leads to changes in hydrodynamics, transport of fines and infilling with sediments (González et al., 2010; Mielck et al., 2019). The sediment characteristics may be altered at the location itself, in its vicinity or in areas further away. Indirect effects of sand extraction result from increased suspended sediment concentrations, which possibly lead to changes in the timing and magnitude of primary production, and resulting changes in food availability for shellfish, fish and birds (van Dalfsen et al., 2000; Boyd et al., 2005; Harezlak et al., 2012). Studies suggest that the dredged areas where sediment is removed superficially (0–2 m deep) recolonise rapidly, with the restoration of biomass and communities to pre-dredge levels anticipated to occur within 2–7 years (van Dalfsen and Essink, 2001; Boyd et al., 2014). In contrast, large-scale extraction will lead to long-lasting changes in morphology, and therefore more permanent changes in faunal communities are expected, for instance on the seascape for fish (Stelzenmüller et al., 2010; Nagelkerken et al., 2015).

Towards More Sustainable Marine Sand Extractions

Through sand extraction, a large number of benthic organisms suffer from mortality. It is not easy to overcome this, but what can be altered are the timing, locations, dimensions, recurrence times, and mining depths of extractions. The depth to which the seafloor is excavated determines the surface area of the impact. Extracting the same volume of sand with a shallow mining depth, down to 2 m below the seafloor, will impact a larger surface area than a deep mining depth of say 6 m below the seafloor. However, it is more likely that deep mining will result in longer-lasting changes in the morphology and sediment composition of the seafloor and therefore the habitats of marine benthic fauna. This greatly depends on local flow conditions and sediment transport, which depends on the orientation of an extraction gully with respect to the tidal flows. The orientation, depth, and slopes of an extraction gully determine altered hydrodynamic patterns, grain sizes, and oxygen levels in deep mining pits. These variables might also affect the transport of fish larvae en route to their nursery rooms. It is, therefore of paramount importance to make detailed hydrodynamic and morphodynamic designs for future mining areas and to develop ecosystem-based design rules for marine extraction sites such as those by de Jong et al. (2016).

Another approach is to work towards ecologically enriched extraction sites. A new physical layout of the seafloor, meaning deeper waters and different currents and sediment characteristics, might offer conditions to develop an enriched ecosystem or even a sanctuary for certain fish species (Rijks et al., 2014). This potential has been tested in a full-scale pilot project in an extraction site in the North Sea in the Building with Nature research programme (de Jong et al., 2015). For the seaward extension of the Port of Rotterdam in the Netherlands, approximately 220 million m^3 of sand was extracted between 2009 and 2013. In order to decrease the surface area of direct impact, the authorities permitted deep sand extraction down to 20 m below the seabed. Two sandbars were artificially created inside the mining area by selective dredging, copying naturally occurring mesoscale bedforms (Baptist et al., 2006), to increase habitat heterogeneity so to increase post-dredging benthic and demersal fish species richness and biomass, Fig. 2. Significant differences in benthos and demersal fish species assemblages in the sand extraction site were found, associated with variables such as water depth, median grain size, fraction of very fine sand and time after the cessation of sand extraction. One and two years after cessation, a significant 20-fold increase in demersal fish biomass was observed in deep parts of the extraction site. In the troughs of the landscaped sandbars; however, a significant drop in fish biomass down to reference levels and a significant change in species assemblage was observed two years after cessation. The fish assemblage at the crests of the sandbars differed significantly from the troughs. This experiment was a first indication of the applicability of landscaping techniques to induce heterogeneity of the seafloor after sand extraction (de Jong et al., 2014).

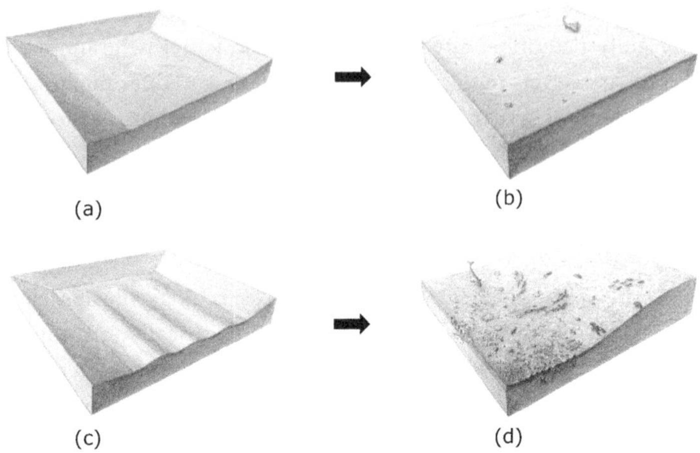

Fig. 2. Conceptualised Images for Seabed Landscaping in Sand Extraction Areas. Landscaped Mining Areas (c) Promote Higher Biodiversity and Productivity (d) After Completion, Compared with Extraction Areas with a Flat Bed (a), Which Yield a Uniform and Poorer Habitat (b) After Sand Extraction. Drawings by Jeroen Helmer/ARK Rewilding Netherlands.

Effects of Sand Nourishments on Life Below Water

A review study by Baptist et al. (2009) described possible ecosystem effects of nourishment. Direct effects are mostly related to the burial of benthic species, but also include effects of increased turbidity on sensitive species (Speybroeck et al., 2006). Burial of benthos has the potential to affect life higher in the food chain, such as shellfish eating birds (Baptist and Leopold, 2009) and increased turbidity affects the foraging success of terns (Baptist and Leopold, 2010). A change in habitat causes indirect effects through the introduction of 'exotic' sediment (i.e. sediment from another location with different properties). Altered sediment properties affect the habitat suitability for benthos, such as penetrability, organic matter content, grain size, and silt content. In testing the habitat selection of juvenile sole, Post et al. (2017) found a clear preference for finer grain sizes, irrespective of temperature, indicating that habitat alteration can affect juvenile survival and subsequently recruitment to adult populations.

Besides the physical effects, the chemical effects of the anaerobic sediment, often together with high sulphide concentrations, play a role. A decreased dissolved oxygen level can amplify the effects of increased sedimentation. Hypoxia (i.e. a lack of oxygen) degrades bottom habitat through a wide suite of mechanisms. Under conditions of limited oxygen at the bottom, rates of nitrogen (nitrate) and phosphate remineralisation, and sulphate reduction increase. The resulting production of nitrite, ammonia, and sulphide in combination with low oxygen can be lethal to benthic organisms (Buzzelli et al., 2002).

Survival, migration, and recruitment all contribute to the recovery after burial by sand nourishments (van Dalfsen and Essink, 2001). The recovery after nourishment depends on many factors, such as the application method/location, the sediment characteristics, the species resistance and resilience and the season of application. Recovery of opportunistic species can sometimes be fairly rapid (e.g. some months to <1 year, because of the quick dispersal of sediments and/or the intrinsic tolerance of the assemblages). Full recovery to community with all species in all life classes can quite often be long-lasting, particularly when the sediments alter the native habitat characteristics, or have high organic loads and/or are highly polluted (Colosio et al., 2007).

Towards More Sustainable Marine Sand Nourishments

To minimise the detrimental effects of sand nourishments, several options are open. The timing and location of sand nourishments determine the ecological effects to a large extent. A mitigating measure is to nourish from October to May, when fish nurseries are not inhabited by 0-year juveniles and outside the main reproduction season of many benthic species. Regarding the location in the offshore profile, the more shallower depths are preferred over the deeper zones, because of the lower diversity and lower biomass of benthos and fish at shallower depths. Of course, the nourished sediment grain sizes should match the local conditions as much as possible.

Technical measures can be taken as well. The risk of burial of benthic species can be greatly reduced when a method is developed that spreads out the nourished sand as a thin layer over a large surface, like salt from a salt shaker. This might be achieved by only slightly opening the bottom doors of a trailing suction hopper dredger while slowly navigating over the nourishment area. To date (2024), such a method has not been tested.

In many circumstances, a temporary alteration of the local morphology following sand nourishment is inevitable. It can then also be considered to maximise seafloor heterogeneity, which possibly leads to an increase in biodiversity. In that case, it is recommended to adhere the nourishment design to existing local morphology, such as spits, bays, or bars. For instance, a shore with parallel breaker bars might offer the possibility to enhance these features. The bars can be made into elongated narrow islands, thus tripling the shoreline and associated shallow coastal habitats. Fine sediment might accumulate in between the bars, thus improving habitat suitability for sole, a valuable flatfish, both ecologically and economically. When these islands subsequently slowly erode due to wind and waves, many temporal and gradual microhabitats will evolve, from which marine biodiversity might profit.

A superlative form of adjusting local morphology has been tested in the Netherlands in the form of a mega-nourishment, which was compared with beach and shoreface nourishments, Table 1.

Table 1. Key Characteristics of a Typical Beach Nourishment, Shoreface Nourishment, and Mega-Nourishment.

Parameter	Beach Nourishment	Shoreface Nourishment	Mega-Nourishment
Volume (10^6 m^3)	1	1	20
Along-shore length (km)	2.5	2.5	2
Intensity (10^6 m^3/km)	0.4	0.4	10
Cross-shore width (km)	0.2	0.5	1
Area (ha)	50	125	200
Thickness (m)	2	0.8	10
Water depth (m)	−1 to +3	−5 to −8	0 to −10
Lifetime (y)	2 to 8	2 to 8	20 to 30

In a pilot experiment named 'Sand Motor' or 'Sand Engine', 21.5 million m^3 of sand was placed at the coast in the shape of a beach spit of 2-km wide and stretching out 1 km into the sea. The expectation was that this enormous volume of sand will protect 10–20 km of coastline over a 20-year period (Stive et al., 2013). The hypothesis was that the environmental pressure of applying a single mega-nourishment is smaller than that of multiple smaller beach or shoreface nourishments over the same period. Evaluating the multi-year development of benthic life Herman et al. (2021) concluded that the Sand Motor has transformed the coast from a linear coastline to a mosaic of underwater landscapes. The enlarged heterogeneity has increased the biodiversity of life below water.

SUMMARY, CONCLUSIONS, AND FUTURE WORK

There are a variety of ways to modify hard coastal structures to increase biodiversity. Most successful adaptations included increasing surface complexity and thus the heterogeneity of living conditions on structures. It should be noted that hard substrates might be non-natural and that the proliferate life on them can compete with local organisms, essentially robbing marine ecosystems of their productivity (Malerba et al., 2019). The warning of Firth et al. (2020) should be repeated that greening of grey infrastructure should not be used to facilitate a non-sustainable coastal development. Habitat alteration of soft substrates may affect ecosystem functioning by interfering with connectivity, leading to the displacement and fragmentation of marine species (Firth et al., 2014; Bishop et al., 2017). Moreover, it has been noticed in recent years that the general lack of structural complexity on artificial surfaces limits epibiotic establishment and therefore overall biodiversity (Lawrence et al., 2021). Additionally, when the diversity of the novel habitat is imbalanced, this can encourage the establishment and proliferation of invasive species (Ruiz et al., 2009; Airoldi and Bulleri, 2011; Dafforn, 2017). These species are detrimental to local marine species and can cause problems for society, such as fouling or reducing fish stocks (Airoldi et al., 2015).

Underwater landscaping through sand extraction should also come with a warning. Although past and present geological and morphological processes have shaped a variety of underwater landscapes and sediment types in the North Sea ranging from gravel beds, via sand waves to mud bottoms, it is preferred to have the least possible disturbance to the seabed instead of the creation of desired habitats. The challenge lies in designing sand extraction pits and gullies that resemble local conditions most and that can recover from the disturbance of mining. In October 2022, a 5 MEuro project with a duration of five years started, from which Operational Recommendations for Ecosystem-based Large-scale Sand Extraction (OR ELSE) will result, https://or-else.nl/, Fig. 3. The project holds positions for 6 PhD students and 3 postdocs working at diverse Dutch universities and has involvement of two Universities of Applied Sciences, as well as Rijkswaterstaat, the dredging industry, research institutes, NGOs, and fishery organisations.

In regard to the recovery from disturbance of sand nourishments, Baptist and Wiersinga (2012) compared different combinations of regular nourishments and mega-nourishments needed to defend the entire 350-km coastline. They showed that for nourishment volumes exceeding 50 million m^3/yr the recurrence time of regular

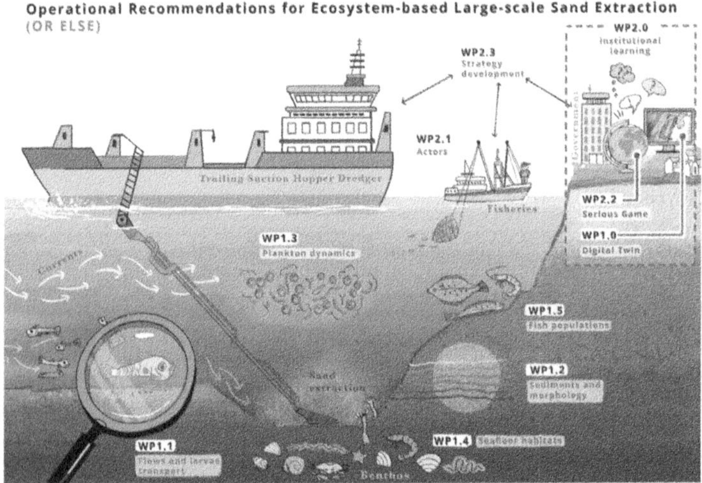

Fig. 3. Conceptual Drawing of a Research Programme on Sustainable Sand Extraction. Drawing by Malou Zuidema.

nourishments along the coast will exceed the recovery period of life below water. For such large volumes, it will be preferable to implement mega-nourishments that reduce recurrence times and defend tens of kilometres of coastline.

In conclusion, enhanced hard coastal defences, underwater landscaping as well as mega-nourishments can lead to more heterogeneous living conditions than at present and although this might lead to an increase in biodiversity, from the viewpoint of nature conservation it is not automatically positive when a measure leads to a change in habitat conditions for life below water. These, and many more lessons, are taught in courses and lectures at universities in the Netherlands, where students at all levels (BSc, MSc, and PhD) have been participating in research and application of Nature-based Solutions. One such example can be found at Wageningen University and Research, which started an entirely new BSc programme 'Marine Sciences' that integrates social and ecological knowledge in Nature-based Solutions to understand, analyse, and provide solutions for the main challenges facing seas, oceans, and coastal regions. Such a multi- and interdisciplinary approach is helping to achieve the goals of SDG14 in higher education. The objective of the BSc Marine Sciences, which runs from September 2023, is to educate internationally oriented bachelor students with interdisciplinary knowledge and skills on marine socio-ecological systems. Graduates will be able to carry out scientific research on the marine social, economic, biological, and biophysical processes. Additionally, evaluating and designing innovative and integrated solutions for the challenges faced in marine socio-ecological systems will contribute in achieving the SDG14 targets and other global initiatives. The programme will educate students to contribute to the exploration, design and development of Nature-based Solutions that are both innovative and inclusive. Based on the extensive expertise at WUR, Nature-based Solutions will be introduced to students as a means of including people and nature to achieve food and energy production, marine nature conservation and coastal defense. Students will be challenged to assess the inclusiveness and sustainability of these Nature-based Solutions against various SDGs, such as zero hunger, climate action, reduced inequality, clean water, and life below water. Educating a new generation of students to become reflective and innovative thinkers that collaborate, cross boundaries and act as agents of change in their future careers will contribute to achieving the goal of SDG14.

ACKNOWLEDGEMENTS

The author wishes to thank all students who have worked with him on Building with Nature, Nature-based Solutions and related topics, and particularly MSc students Sjoerd de Jong, Jack Grehan and Arjen Schokker for studying hard substrate fauna on sea defences during their internship at BAM, and PhD students Maarten de Jong and Marjolein Post for studying sand extractions and sand nourishments. The author further thanks Jan Philipsen and Jetske ten Caat for their leadership and guidance in developing the new BSc Marine Sciences at WUR.

REFERENCES

Airoldi, L., Turon, X., Perkol-Finkel, S., Rius, M., St, N., Aviv, T. and Science, E. 2015. Corridors for aliens but not for natives: effects of marine urban sprawl at a regional scale, *Diversity and Distributions*, 21(7), 755–768. doi: 10.1111/DDI.12301

Airoldi, L. and Bulleri, F. 2011. Anthropogenic disturbance can determine the magnitude of opportunistic species responses on marine urban infrastructures, *PLOS ONE*, 6(8), e22985. doi: 10.1371/JOURNAL.PONE.0022985

Baptist, M., Van der Meer, J. and De Vries, M. 2007. *De Rijke Dijk, ontwerp en benutting van harde infrastructuur in de getijzone voor ecologische en recreatieve waarden*, Delft, Delft University of Technology. Available at: http://www.innover enmetwater.nl/upload/documents/Haalbaarheidsstudie De Rijke Dijk 2006 (rapport).pdf

Baptist, M.J., van Dalfsen, J., Weber, A., Passchier, S. and van Heteren, S. 2006. The distribution of macrozoobenthos in the southern North Sea in relation to meso-scale bedforms, *Estuarine, Coastal and Shelf Science*, 68(3–4), 538–546. doi: 10.1016/j.ecss.2006.02.023

Baptist, M.J., Tamis, J.E., Borsje, B.W. and van der Werf, J.J. 2009. *Review of the Geomorphological, Benthic Ecological and Biogeomorphological Effects of Nourishments on the Shoreface and Surf Zone of the Dutch Coast.* IMARES C113/08, Deltares Z4582.50. Available at: http://edepot.wur.nl/8938

Baptist, M.J. and Leopold, M.F. 2009. The effects of shoreface nourishments on Spisula and scoters in The Netherlands, *Marine Environmental Research*, 68(1), 1–11. doi: 10.1016/j.marenvres.2009.03.003

Baptist, M.J. and Leopold, M.F. 2010. Prey capture success of Sandwich Terns Sterna sandvicensis varies non-linearly with water transparency, *Ibis*, 152(4), 815–825. doi: 10.1111/j.1474-919X.2010.01054.x

Baptist, M.J. and Wiersinga, W. 2012. Zand erover. Vier scenario's voor zachte kustverdediging, *De levende natuur*, 113(2), 56–61.

Bishop, M.J., Mayer-Pinto, M., Airoldi, L., Firth, L.B., Morris, R.L., Loke, L.H.L., Hawkins, S.J., et al. 2017. Effects of ocean sprawl on ecological connectivity: impacts and solutions, *Journal of Experimental Marine Biology and Ecology*, 492, 7–30. doi: 10.1016/J.JEMBE.2017.01.021

Boyd, S.E., Limpenny, D.S., Rees, H.L., Cooper, K.M., et al. 2005. The effects of marine sand and gravel extraction on the macrobenthos at a commercial dredging site (results 6 years post-dredging), *ICES Journal of Marine Science*, 62, 145–162. doi: 10.1016/j.icesjms.2004.11.014

Boyd, S.E., Cooper, K.M., Limpenny, D.S., Kilbride, R., Rees, H.L., Dearnaley, M.P., Stevenson, J., et al. 2014. *Assessment of the Re-habilitation of the Seabed Following Marine Aggregate Dredging*, Science Series Technical Report, Centre for Environment, Fisheries and Aquaculture Science, 121. Available at: https://www.researchgate.net/publication/228915400 [Accessed 9 April 2020].

Buzzelli, C.P., Luettich, R.A., Powers, S.P., Peterson, C.H., McNinch, J.E., Pinckney, J.L. and Paerl, H.W. 2002. Estimating the spatial extent of bottom-water hypoxia and habitat degradation in a shallow estuary, *Marine Ecology Progress Series*, 230, 103–112. doi: 10.3354/MEPS230103

Chapman, M.G. and Underwood, A.J. 2011. Evaluation of ecological engineering of "armoured" shorelines to improve their value as habitat,

Journal of Experimental Marine Biology and Ecology, 302–313. doi: 10.1016/j.jembe.2011.02.025

Colosio, F., Abbiati, M. and Airoldi, L. 2007. Effects of beach nourishment on sediments and benthic assemblages, *Marine Pollution Bulletin*, 54(8), 1197–1206. doi: 10.1016/J.MARPOLBUL.2007.04.007

Dafforn, K.A. 2017. Eco-engineering and management strategies for marine infrastructure to reduce establishment and dispersal of non-indigenous species, *Management of Biological Invasions*, 8(2), 153–161. doi: 10.3391/mbi.2017.8.2.03

van Dalfsen, J.A., Essink, K., Madsen, H.T., Birklund, J., Romero, J. and Manzanera, M. 2000. Differential response of macrozoobenthos to marine sand extraction in the North Sea and the Western Mediterranean, *ICES Journal of Marine Science*, 57, 1439–1445. doi: 10.1006/jmsc.2000.0919

van Dalfsen, J.A. and Essink, K. 2001. Benthic community response to sand dredging and shoreface nourishment in Dutch coastal waters, *Senckenbergiana maritima*, 31(2), 329–332.

Ellis, J.C., Shulman, M.J., Wood, M., Witman, J.D. and Lozyniak, S. 2007. Regulation of intertidal food webs by avian predators on New England rocky shores, *Ecology*, 88(4), 853–863. doi: 10.1890/06-0593

Firth, L.B., Thompson, R.C., Bohn, K., Abbiati, M., Airoldi, L., Bouma, T.J., Bozzeda, F., et al. 2014. Between a rock and a hard place: Environmental and engineering considerations when designing coastal defence structures, *Coastal Engineering*, 87, 122–135. doi: 10.1016/J.COASTALENG.2013.10.015

Firth, L.B., Airoldi, L., Bulleri, F., Challinor, S., Chee, S.-Y., Evans, A.J., Hanley, M.E., et al. 2020. Greening of grey infrastructure should not be used as a Trojan horse to facilitate coastal development, *Journal of Applied Ecology*, 0–2. doi: 10.1111/1365-2664.13683

González, M., Medina, R., Espejo, A., Tintoré, J., Martin, D. and Orfila, A. 2010. Morphodynamic evolution of dredged sandpits, *Journal of Coastal Research*, 263(263), 485–502. doi: 10.2112/08-1034.1

Haasnoot, M., Haasnoot, M., Bouwer, L., Diermanse, F., Kwadijk, J., Van der Spek, A., Oude Essink, G., Delsman, J., et al. 2018. *Mogelijke gevolgen van versnelde zeespiegelstijging voor het Deltaprogramma. Een verkenning*, Delft, Deltares rapport 11202230-005-0002.

Harezlak, V., van Rooijen, A., Friocourt, Y., van Kessel, T. and Los, H. 2012. *Winning suppletiezand Noordzee*, Delft, Deltares 1204963-000.

Herman, P.M.J., Moons, J.J.S., Wijsman, J.W.M., Luijendijk, A.P. and Ysebaert, T. 2021. A mega-nourishment (sand motor) affects landscape diversity of Subtidal Benthic Fauna, *Frontiers in Marine Science*, 8. doi: 10.3389/fmars.2021.643674

Ido, S. and Shimrit, P.F. 2015. Blue is the new green – ecological enhancement of concrete based coastal and marine infrastructure, *Ecological Engineering*, 84, 260–272. doi: 10.1016/J.ECOLENG. 2015.09.016

de Jong, M.F., Baptist, M.J., van Hal, R., de Boois, I.J., Lindeboom, H.J. and Hoekstra, P. 2014. Impact on demersal fish of a large-scale and deep sand extraction site with ecosystem-based landscaped sandbars, *Estuarine, Coastal and Shelf Science*, 146, 83–94. doi: 10.1016/ j.ecss.2014.05.029

de Jong, M.F., Baptist, M.J., Lindeboom, H.J. and Hoekstra, P. 2015. Short-term impact of deep sand extraction and ecosystem-based landscaping on macrozoobenthos and sediment characteristics, *Marine Pollution Bulletin*, 97(1–2). doi: 10.1016/j.marpolbul.2015.06.002

de Jong, M.F., Borsje, B.W., Baptist, M.J., van der Wal, J.T., Lindeboom, H.J. and Hoekstra, P. 2016. Ecosystem-based design rules for marine sand extraction sites, *Ecological Engineering*, 87. doi: 10.1016/j.ecoleng. 2015.11.053

Lawrence, P.J., Evans, A.J., Jackson-Bué, T., Brooks, P.R., Crowe, T.P., Dozier, A.E., Jenkins, S.R., et al. 2021. Artificial shorelines lack natural structural complexity across scales, *Proceedings of the Royal Society B*, 288(1951). doi: 10.1098/RSPB.2021.0329

Lodder, Q.J., Slinger, J.H., Wang, Z.B. and van Gelder, C. 2019. Decision making in Dutch coastal research based on coastal management policy assumptions. In *Coastal Management 2019: Joining Forces to Shape Our Future Coasts*, pp. 291–300, ICE.

Malerba, M.E., White, C.R. and Marshall, D.J. 2019. The outsized trophic footprint of marine urbanization, *Frontiers in Ecology and the Environment*, 17(7), 400–406. doi: 10.1002/fee.2074

Mendonça, V., Madeira, C., Dias, M., Vermandele, F., Archambault, P., Dissanayake, A., Canning-Clode, J., et al. 2018. What's in a tide pool? Just as much food web network complexity as in large open ecosystems, *PLoS ONE*, 13(7). doi: 10.1371/journal.pone.0200066

Mielck, F., Hass, H.C., Michaelis, R., Sander, L., Papenmeier, S. and Wiltshire, K.H. 2019. Morphological changes due to marine aggregate extraction for beach nourishment in the German Bight (SE North Sea), *Geo-Marine Letters*, 39(1), 47–58. doi: 10.1007/s00367-018-0556-4

Nagelkerken, I., Sheaves, M., Baker, R. and Connolly, R.M. 2015. The seascape nursery: a novel spatial approach to identify and manage nurseries for coastal marine fauna, *Fish and Fisheries*, 16(2), 362–371. doi: 10.1111/faf.12057

Post, M.H.M., Blom, E., Chen, C., Bolle, L.J. and Baptist, M.J. 2017. Habitat selection of juvenile sole (Solea solea L.): Consequences for shoreface nourishment, *Journal of Sea Research*, 122. doi: 10.1016/j.seares.2017.02.011

Rijks, D.C., de Jong, M.F., Baptist, M.J. and Aarninkhof, S.G.J. 2014. Utilising the full potential of dredging works: ecologically enriched extraction sites, *Terra et Aqua*, 136, 5–15.

Ruiz, G.M., Freestone, A.L., Fofonoff, P.W. and Simkanin, C. 2009. Habitat distribution and heterogeneity in marine invasion dynamics: the importance of hard substrate and artificial structure, 321–332. doi: 10.1007/B76710_23

Speybroeck, J., Bonte, D., Courtens, W., Gheskiere, T., Grootaert, P., Maelfait, J.P., Mathys, M., et al. 2006. Beach nourishment: an ecologically sound coastal defence alternative? A review, *Aquatic Conservation: Marine and Freshwater Ecosystems*, 16(4), 419–435. doi: 10.1002/AQC.733

Stelzenmüller, V., Ellis, J.R. and Rogers, S.I. 2010. Towards a spatially explicit risk assessment for marine management: Assessing the vulnerability of fish to aggregate extraction, *Biological Conservation*, 143(1), 230–238. doi: 10.1016/j.biocon.2009.10.007

Stive, M.J., de Schipper, M.A., Luijendijk, A.P., Aarninkhof, S.G., van Gelder-Maas, C., van Thiel de Vries, J.S., de Vries, S., et al. 2013. A new alternative to saving our beaches from sea-level rise: the sand engine, *Journal of Coastal Research*, 29(5), 1001–1008.

Strain, E.M.A., Olabarria, C., Mayer-Pinto, M., Cumbo, V., Morris, R.L., Bugnot, A.B., Dafforn, K.A., et al. 2018. Eco-engineering urban infrastructure for marine and coastal biodiversity: Which interventions have the greatest ecological benefit?, *Journal of Applied Ecology*, 55(1), 426–441. doi: 10.1111/1365-2664.12961

de Vriend, H.J. and van Koningsveld, M. 2012. *Building with Nature: Thinking, Acting and Interacting Differently*, Dordrecht, EcoShape, The Netherlands.

5

SEAWEEDS FROM AN ITALIAN COASTAL LAGOON: FROM 'WASTE MATERIAL' TO COMMERCIALLY VALUABLE MARINE RESOURCES

Caterina Pezzola

University of Groningen, Groningen, Netherlands

ABSTRACT

Marine seaweeds, characterised by high-valued bioactive compounds, are used worldwide for several applications, including human food, animal feed, pharmaceutics and cosmetics, bioplastics, agricultural fertilisers, biofuels, and others. Seaweed production can be carried out through different approaches, from on-land or sea-based cultivation to the harvesting of wild stocks. The latter can be of particular importance in the case of seasonal algal over-proliferations, often caused by eutrophic conditions associated with intensive human industrial activities, and which wreak havoc with ecosystem functioning and hinder economic activities. In Europe, Italy experiences seaweed blooms in several coastal basins, such as the Lagoon of Venice and the Lagoon of Orbetello (Tuscany). Here, the proliferating seaweed represents a disturbance to the natural ecosystem and to local business and touristic activities. These biomasses hold no economic value in the country and are systemically removed and disposed of. Re-purposing the biomass to produce seaweed-derived commercial goods would provide benefits for the environment and local economic activities while promoting a sustainable business within a Circular Economy framework and contribute to the UN Sustainable Development Goals number 12 ('Responsible consumption and production'), and number 14 ('Life under water'), among others.

Keywords: Seaweed; circular economy; algal blooms; recycling; biorefinery; Italy

INTRODUCTION

Seaweeds

Macroscopic marine algae, commonly known as 'seaweed', play a key role in ecosystem functioning. Seaweeds provide a wide range of ecosystem services (ESs), including the preservation of marine biodiversity, oxygen production, and contribution to particulate and organic matter cycles (Cabral et al., 2016; Lotze et al., 2018). Furthermore, seaweeds can play a central part in climate change mitigation by performing carbon sequestration, regulating eutrophication events, and taking up heavy metals from the water column.

Macroalgae can be divided into three classes: brown, red, and green seaweeds. Every species within each class has different features and characteristics in terms of geographical distribution, growth rates, bioactive compound composition (i.e. proteins, minerals, pigments, sugars, etc. content), and more. Class- and species-specific features are also influenced by environmental factors such as solar irradiation, water temperatures, salinity rates, and nutrient availability (Makkar and Tran, 2016).

Seaweed Market

According to the Food and Agriculture Organization (FAO), the seaweed market is globally on an upward trajectory and is expected to grow exponentially in the coming years. Worldwide, seaweed production is largely carried out in two different ways: (1) on-land or sea-based cultivation, also referred to as 'seaweed farming', or (2) harvesting of wild stocks. In 2019, seaweed farming contributed to nearly 30% of global aquaculture production, including the established industries of finfish and shellfish aquaculture (FAO, 2024). The bulk of global seaweed aquaculture production comes from East and Southeast Asian countries, where seaweeds play a prominent role in traditional human diets (FAO, 2018). In Western countries, seaweed aquaculture is at a very early stage of development and seaweed-derived products are still niche and overlooked by the public.

In recent years, given the multiple environmental and industrial benefits linked to the seaweed sector, Western countries are increasingly turning towards commercial production of seaweed and exploring novel applications for this sustainable marine crop. This positive trend is particularly driven by a rising public perception of the negative environmental impact of traditional food production systems and by a growing market demand for healthier and more sustainable food alternatives.

Seaweed Applicability

Seaweed could meet customers' demands due to its novel compound composition and therapeutical properties. The concentration of nutrients in seaweeds is generally 10–20 times higher compared to terrestrial plants, due to the facilitated absorption of elements from seawater (Moreda-Piñeiro et al., 2012). Most seaweeds are low in lipids (1–5%) and high in protein and polysaccharides (Terasaki et al., 2009; Ito et al., 2018). Due to their high-valued and novel bioactive compound composition, seaweeds are being explored worldwide for several industrial applications, other than human food. These include animal feed, pharmaceutics and cosmetics products, feedstock in biorefineries (e.g. to produce biofuels), agricultural fertilisers, bioplastics, and more. Furthermore, due to their capacity to take up heavy metal ions from seawater, seaweed entails high bioremediation and biomonitoring potential. However, due to their easiness in taking up and accumulating nutrients and heavy metals, seaweed-derived food and feed products need to adhere to international thresholds of heavy metals and contaminants [arsenic (As), cadmium (Cd), lead (Pb), chloride (Cl), polychlorinated biphenyls (PCB)] content prior to entrance to the market.

Sustainability Goals

Because of their multiple environmental benefits, seaweeds can significantly contribute to sustainability goals set by the United Nations (UN). The use of seaweeds specifically falls within the framework of Sustainable Development Goal 14 (SDG14) – Life Under Water. SDG14 focuses on ocean preservation and the use of marine resources to achieve sustainable economic development. Within this context, seaweeds not only play a crucial role in reducing marine pollution (SDG target 14.1) by taking up CO_2 and pollutants, and in increasing the ecosystem's resilience (SDG target 14.2), but also represent a valuable marine resource for commercial purposes (SDG target 14.7). Preserving natural seaweed stocks, supporting the aquaculture sector through financial incentives and favourable policies, and promoting seaweed-related benefits among consumers, are compelling factors for boosting the sector and for the achievement of SDG14 targets.

Seaweed Proliferations

Although seaweeds bring about several benefits, they can also represent a liability for ecosystems and human activities. A few seaweed species, particularly green ones (e.g. *Ulva* sp., *Chaetomorpha* sp.), tend to over-proliferate under given environmental conditions and form the so-called 'macroalgal blooms'. These blooming events are mainly caused by the combination of human industrial activities, leading to eutrophication events, (i.e. excess of nutrients in the water column), and by prolonged heat during the summertime. Seasonal proliferations raise several environmental and societal concerns as they wreak havoc with marine ecosystems functioning (e.g. fish die-offs) and touristic and fishery activities. Globally, these events have been increasing in frequency and severity (Ye et al., 2011).

Southern Europe, and specifically the Mediterranean basin, can count on favourable climate conditions for the growth of large volumes of seaweeds. Among Mediterranean countries, Italy often experiences over-proliferations of green seaweeds. Macroalgal blooms of the green algae *Ulva* spp. have been recorded in the North Adriatic Sea and the Lagoon of Venice and represented a big economic and societal issue, particularly during the late 1980s and early 1990s. Nowadays, the overgrowth of the green seaweed *Chaetomorpha linum* is recorded in the lagoon of Orbetello, on the west coast of the country.

In Italy, the over-proliferating biomasses are generally harvested by local authorities to avoid environmental and economic disruptions. The harvested material holds little or no economic value for local communities, uneducated about the benefits of seaweeds, and is thus commonly discarded as waste. However, within a Circular Economy framework, harvesting bloom-forming seaweeds to then re-use them for commercial purposes can be a cost-efficient way to control their spreading and mitigate harmful effects (Dominguez and Loret, 2019), while contributing to the valorisation of a novel and sustainable source of bioactive compounds.

Due to their ability to uptake and store nutrients and metals, seaweeds colonising eutrophic, nutrients-rich environments can display high levels of contaminants such as polychlorinated biphenyls (PCBs), arsenic (As), cadmium (Cd), mercury (Hg), lead (Pb), which are toxic and/or carcinogenic for humans and animals (IARC, 1987). The concentration of these elements in seaweed tissues can strongly limit the commercial use of these biomasses, particularly for products destined for the pharmaceuticals, cosmetics, human food, and animal feed industries. In recent years, the development of novel and advanced biorefinery technologies allow for the selection and extraction of specific compounds out of the algal matrix, while discarding the contaminated parts. These technologies would allow for the application of contaminated seaweeds in a broad range of industries.

The Lagoon of Orbetello-Case Study

The lagoon of Orbetello, located along the Tyrrhenian coast in the Tuscany region (Italy), covers an area of 25.25 km^2 and is made of two communicating basins (Fig. 1). The basin represents a site of high environmental and cultural interest, hosting a Natural WWF Oasis, and holds a crucial economic role due to the presence of several aquaculture sites, fishery activities, and more.

The intense human activity around the basin has led to hypertrophic conditions in the lagoon of Orbetello. The continuous dumping of urban, aquaculture, and agricultural wastewater into the basins has led to major macroalgal proliferation (up to 24 kg m^{-2}). Algal biomass is produced almost constantly throughout the year and blooms are mainly composed of the opportunistic, mats-forming, filamentous seaweed *Chaetomorpha linum* (Müller) Kützing (Lenzi et al., 2013; McGlathery et al., 1996) (Fig. 1).

C. linum thrives in eutrophic environments and can be particularly resistant to adverse environmental conditions (Lenzi et al., 2017). Due to locally mild winter temperatures, *C. linum* mats can

Fig. 1. Lump of the Green Filamentous Seaweed *Chaetomorpha linum* from the *Orbetello lagoon.*

grow at the surface almost all year round. Dead algal thalli, decomposed by sulfate-reducing bacteria, sink to the muddy bottom of the lagoon and release nutrients (Fig. 2). This cycle allows the seaweed to persist all year round within a self-sustainment process (Krause-Jensen et al., 1996). The activity of sulfate-reducing bacteria, combined with the constant presence of *C. linum*, increasingly lead to dystrophic events wreaking havoc with the lacunar ecosystem (e.g. fish-die-off events in particularly hot summers) (Lenzi & Renzi, 2011). Henceforth, the algal biomass represents a great environmental and economic concern for regional authorities and for the local community. Proliferations can however be prevented or offset through the harvesting of the material (Lenzi, 1992; De Leo et al., 2002).

The lagoon's seaweed biomass is currently classified as 'urban waste', according to national legislation. In compliance with national legislative guidelines regarding waste management, the material is periodically removed and disposed of by the municipality of Orbetello. For the parties involved, removal and disposition of seaweeds entail high costs (approximately 170 €/ton/year),

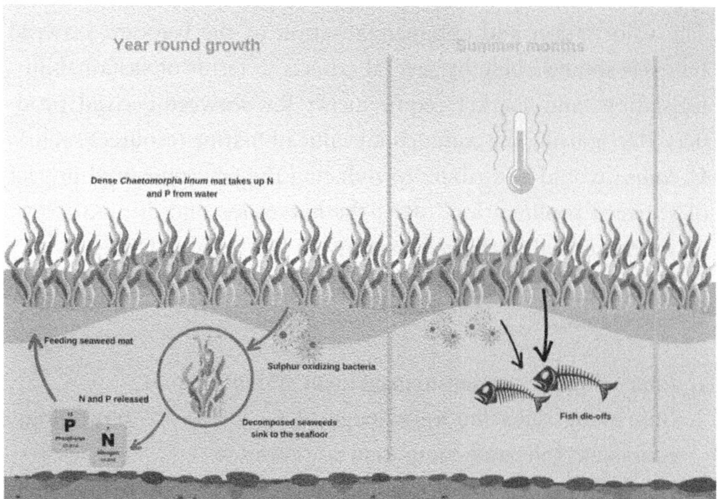

Fig. 2. Pathways Linked to the Year-Round Growth of *C. linum* Mats at Surface and Its Association with Sulphur-Oxidising Bacteria in the Lagoon of Orbetello.

which are not offset by any revenues. Moreover, the state-of-the-art available harvesting equipment, together with the inadequate funds allocated to waste management, keep the removal capacity low. Ongoing harvesting operations do not represent a sufficient remedy to the environmental issues tied to the year-round presence of C. *linum* and its harmful impact on the ecosystem. The current harvesting capacity fluctuates around 1,000 tonnes (fresh weight) per year, while it is estimated that the C. *linum* biomass reaches a constant volume of up to 100,000 tonnes.

A process of 'waste valorisation' (Idowu et al., 2013), aimed at recycling and repurposing the material harvested from the lagoon into valuable industrial products or as a source of energy, would provide an efficient and sustainable solution to exploit natural resources within a Circular Economy framework. Through industrial biorefinery processes, the raw algal material can be converted into a wide range of by-products, from a source of vegan protein for human and animal diets to a crop for green energy production.

SCOPE

The valorisation and commercialisation of the lagoon's seaweed biomass should abide by several criteria in terms of sustainability, legislation, and market requirements for seaweed-derived products. Recognising the commercial value of marine resources such as C. *linum* would contribute to reducing the environmental impact of seaweed proliferation, offset the harvesting and disposal costs, and meet several of the goals set by the UN within SDG14.

Valorising the lagoon of Orbetello and its resources entails several stages:

1. Improving the current management system (Fig. 3)
2. Improving scientific knowledge of C. *linum* eco-physiology, ecosystem role, and compound composition
3. Determining C. *linum* applicability and potential based on quality thresholds and bioactive compound composition
4. Creating a value chain within a sustainable Circular Economy framework (Fig. 3).

These objectives are in line with targets 14.1 ('Decrease marine pollution'), 14.2 ('Protect and restore ecosystems'), 14.7 ('Increase the economic benefits from the sustainable use of marine resources'), and 14.8 ('Improve scientific knowledge and marine technologies') of UN SDG14.

APPROACH

Improving the Current Management System

In compliance with national and regional legislative guidelines, algal biomass is classified and treated as either 'urban waste' or 'special waste' and is thus allocated to disposal. Current harvesting operations entail local fishermen collecting material weekly, almost all year round, employing boats owned by the local administration. Thereafter, the material brought onshore is gathered and transported to a disposal facility in the area (Fig. 3). Such a management system calls for considerable investments from all parties involved (i.e. municipality, Region Toscana, local community).

As the high costs are coupled with the limited funds allocated to waste management in the region, no investments can be made to

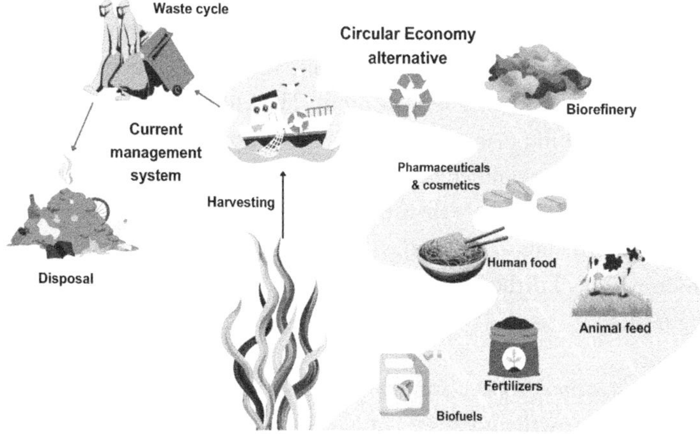

Fig. 3. Infographic of the Current Management System of *C. linum* Carried Out by Local Authorities and the Proposed Alternative System, Based Upon a Circular Economy System.

upgrade or replace the outdated and inadequate harvesting equipment currently deployed by the fishermen. Ongoing removal operations only cover ca. 1% of the available volumes of seaweed in the lagoon. Therefore, environmental issues linked to *C. linum* proliferation are hardly solved by the remediation scheme currently managed by local administrations.

Avoiding the 'waste cycle' and re-allocating the harvested biomass to a Circular Economy scheme would be the first step to producing revenues and offsetting harvesting costs (Fig. 3). As a by-product, the creation of profit could lead to increased social acceptability and support of the capitalisation of the lagoon's resources from the local community.

Finally, the creation of revenues can provide capital for, and economic interest towards, an increase in the current harvesting capacity. Profits can be invested in the renovation or replacement of obsolete harvesting equipment and in the overall improvement of ongoing operations. Ultimately, this can avoid the onset of anoxic and sulfate-reduction events, usually happening during hot summers and expected to increase in frequency and severity due to climate change.

Improving Scientific Knowledge of C. Linum

C. linum has so far been largely understudied, especially compared to commercially available seaweeds like the green *Ulva* spp. or the brown *Saccharina latissima* (Kombu). Therefore, we know little of the ecological importance of this species within the lagoon ecosystem and of the commercial potential it holds within the market. Extending the current scientific knowledge is key to sustainably exploiting the lagoon's biomass and avoiding further harm to the environment. Further research should revolve around the following topics:

1. *The lagoon's trophic status and macronutrient cycles* – To predict the growth and effects of *C. linum* mats on the surrounding environment, it is crucial to identify productivity (trophic) areas within the lagoon and to investigate nutrient cycles in water

and sediments. Algal distribution and density are strongly influenced by factors like nutrient balances and fluxes, physical and chemical stratification of the water column (Krause-Jensen et al., 1996), composition and redox potential of sediments (Lenzi et al., 2013a), benthic uptake of nutrients, and more.

2. *The role of C. linum mats within the lagoon's ecosystem* – Understanding the role of seaweed mats within the ecosystem functioning of the lagoon is crucial to determining the optimal scale and timing of harvesting. Thus, decisions should be taken upon seaweed and algal mats' features such as algal density, the concentration of nutrients within the seaweed, thalli degradation rates, bacterial respiration within the mat, etc. (Lenzi et al., 2013b).
3. *Seasonal changes in bioactive compound composition* – Environmental conditions, such as solar irradiance and water temperature, strongly influence the bioactive composition of seaweeds (Olsson et al., 2020). Seasonal environmental variations affect the seaweeds' uptake of nutrients from the water, their pigment composition, and the internal synthesis of primary and secondary metabolites (Marinho-Soriano et al., 2006). Therefore, seaweed's bioactive composition (e.g. protein, carbohydrates, amino acids, fatty acids, etc.) fluctuates throughout the seasons. Ultimately, internal composition changes not only affect the role of macroalgae in the ecosystem but also play an important part in determining their commercial value. The full range of *C. linum*'s seasonal compound composition variations has not yet been described.

Determining C. Linum Applicability and Potential

Hereafter, the aim is to define the most suitable industrial application for *C. linum* biomass based on:

1. *C. linum* species-specific characteristics and seasonal variations in compound composition
2. Quality of the material based on heavy metals and pollutants accumulated in the algal tissue

Overall, the market applicability of seaweeds can be explained by the bio-cascading pyramid illustrated in Fig. 4. The pyramid highlights the reverse relationship between the 'market value' of the product (i.e. price and quality requirements), and the volumes required for biorefinery processes. For instance: the commercialisation of seaweed-based biofuels is not bound to high-quality standards (i.e. in terms of pollutant concentrations), but the production is challenged by the need for significantly large volumes of raw material (i.e. in the order of thousands of tonnes). Oppositely, the volume requirements for pharmaceuticals or cosmetical products are usually as low as a few kilograms of seaweed extract. However, seaweed biomass intended for these high-value sectors is limited by strict quality thresholds and monitored by mandated agencies.

The pyramid can also be interpreted as a visual conceptualisation of the 'cascading biorefinery' approach, where the side- and waste-stream of the biorefinery of high-value products (e.g. proteins extraction for food and feed) are recycled for lower-value by-products (e.g. biofuels). This approach is aimed at maximising the output of the raw material and adding commercial value to all the compounds found in the biomass (van Hal et al., 2014). Seaweed contains a wide range of different-valued compounds, from

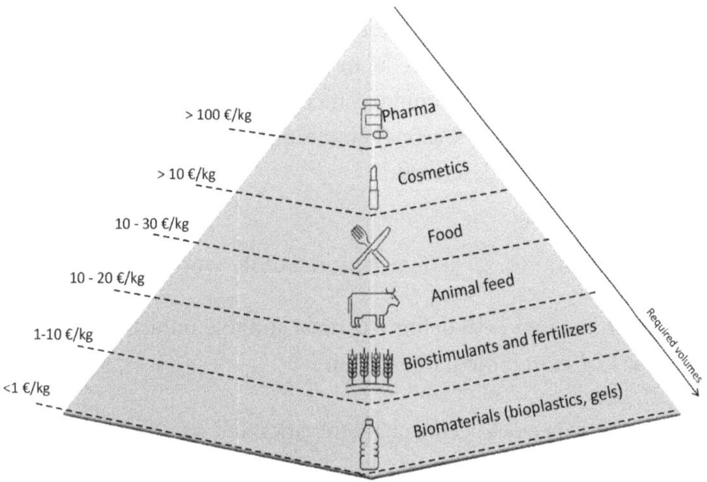

Fig. 4. Bio-Cascading Pyramid Depicting the Inverse Relationship Between Quality/Profitability of a Product and the Amounts Required for Its Production.

high-value proteins, minerals, and pigments for the top-of-the-pyramid sectors, to low-value components (e.g. glucose) for fermentation processes within the energy and biomaterials industries. Therefore, seaweed biomass is considered particularly suitable for the adoption of a cascading biorefinery scheme.

When it comes to optimising production and commercialisation, seasonality is another crucial factor to consider. Bioactive compound composition not only fluctuates due to abiotic and biotic interactions in the environment but is also strongly species-specific. As previously mentioned, *C. linum* is so far understudied compared to other commercially viable seaweeds. Thus, monitoring composition changes by periodically analysing samples of *C. linum* from the lagoon is key to determining changes throughout the year and assigning the biomass a market value.

In the Orbetello lagoon, macroalgal proliferations are triggered by human-induced eutrophication events, indicating high concentrations of pollutants in the water. As seaweeds easily uptake contaminants from the water and accumulate them in their tissues, *C. linum* is expected to be high in heavy metals concentration for direct use in many high-value sectors such as the food industry. Therefore, a thoughtful biorefinery approach should be laid out to exploit both high- and low-value components in *C. linum*, while discarding contaminants. Extraction of compounds from the algal matrix and recycling waste-stream by-products would allow for inclusive and efficient exploitation of the raw material. Several extraction methods could be applied to recover high-value compounds from a low-quality matrix based on the desired final product. The state-of-the-art seaweed industry sees extracted fractions as having a wider commercial potential and applicability range than unprocessed raw material (Wang et al., 2020). The most common extraction techniques involve the use of solvents, heat, infusion, or steam distillation (Bordoloi and Goosen, 2020). However, these extraction approaches are often energy-, time-, and material-consuming, and thus costly. New extraction methods (e.g. supercritical fluid, enzyme- and ultrasound-assisted extraction, etc.) are currently being developed all around the world to make the process cheaper, faster, more efficient, and eco-friendlier (Matos et al., 2021).

Within the lagoon context, in line with the previously mentioned 'cascading biorefinery' approach (Fig. 4), the waste stream generated during the extraction of high-value compounds from *C. linum* is to be directed towards the energy and biofuel sectors. The high volumes available are suitable for the establishment of a biorefinery scheme for low-value products that require considerable volumes of raw material, such as biofuels.

Creating a Value Chain Within a Sustainable Circular Economy Framework

Within the commercial valorisation of *C. linum*, it is essential to prioritise sustainability and the preservation of the aquatic system. While enhancing the current harvesting capacity is expected to benefit the lagoon, seaweeds in the lagoon play a key role in supporting biodiversity and mitigating the eutrophication of seawater by taking up excess nutrients. Therefore, harvesting too large volumes can wreak havoc with ecosystem functioning and the preservation of natural aquatic communities (Lorentsen et al., 2010).

Combining a strong business strategy with sustainability targets (e.g. SDG14) is key to increasing social acceptance and market demand for the final product, and ultimately for the delineation of a successful business case.

To better address the trade-offs associated with the process, from harvesting the raw material to manufacturing and finally commercialising the final product, a Life Cycle Sustainability Assessment (LCSA) should be employed. The LCSA can help predict and evaluate all the potential social, economic, and environmental benefits and/or negative impacts related to every step of the process. This tool can guide decision-makers in choosing the most sustainable technologies and resources to produce cost-efficient, eco-friendly, and socially responsible products, contributing to the UN Sustainable Development Goals.

CONCLUSIONS

Appointing commercial value to the *Chaetomorpha linum* biomass from the Orbetello lagoon offers an efficient and cost-effective solution to the environmental concerns related to seasonal algal proliferation. At the same time, repurposing the material would support a growing European seaweed market and contribute to the valorisation of an overlooked but valuable marine resource. Furthermore, like most seaweeds, the filamentous C. *linum* can grant a flexible business approach, where both high- and low-value compounds in the algal matrix can be exploited within different industrial sectors. The availability of high volumes within the lagoon facilitates the use of waste-stream for by-products, such as biofuels, that require large amounts of raw material stocks. Further research into the species-specific and seasonal composition properties of C. *linum* is thus necessary to determine the optimal cascading biorefinery approach and to establish a solid business case.

Nonetheless, making sustainable use of marine resources for industrial purposes requires a balance between economic interests, scientific understanding, and environmental awareness. Therefore, an LCSA must improve knowledge of C. *linum*'s role within the lagoon ecosystem. Determining the optimal scale and timing of harvesting is crucial to avoid wreaking havoc on the lagoon's ecological equilibrium. Finally, the latest and most eco-friendly biorefinery technologies should be deployed to reduce the environmental impact of seaweed biorefinery processes.

This Circular Economy-based approach would greatly contribute to the achievement of the following UN Sustainable Development Goals 14 targets:

- *Target 14.1* – Reduce marine pollution, particularly from land-based activities, including nutrient pollution
- *Target 14.2* – Protect and restore marine and coastal ecosystems to avoid significant adverse impact
- *Target 14.7* – Increase economic benefits from the sustainable use of marine resources
- *Target 14.8* – Increase scientific knowledge, improve research capacity, and transfer marine technology to preserve ocean health

Finally, increasing social awareness and public demand for seaweed-derived products and developing greener industrial-scale biorefinery technologies are essential factors for, respectively, achieving an economically successful business and meeting sustainability goals throughout the entire production chain, from the lagoon's 'waste' material to commercially valuable products.

REFERENCES

Bordoloi, A. and Goosen, N. 2020. Green and integrated processing approaches for the recovery of high-value compounds from brown seaweeds. In *Advances in Botanical Research*, pp. 369–413, Vol. 95, Academic Press. doi: 10.1016/bs.abr.2019.11.011

Cabral, P., Levrel, H., Viard, F., Frangoudes, K., Girard, S. and Scemama, P. 2016. Ecosystem services assessment and compensation costs for installing seaweed farms, *Marine Policy*, 71, 157–165. doi: 10.1016/j.marpol.2016.05.031

De Leo, G.A., Bartoli, M., Naldi, M. and Viaroli, P. 2002. A first-generation stochastic bioeconomic analysis of algal bloom control in a coastal lagoon (Sacca di Goro, Po River Delta), *Marine Ecology*, 23, 92–100. doi: 10.1111/j.1439-0485.2002.tb00010.x

Dominguez, H. and Loret, E.P. 2019. *Ulva lactuca*, a source of troubles and potential riches, *Marine Drugs*, *17*(6), 357. doi: 10.3390/md17060357

FAO, F. 2018. Agriculture Organization of the United Nations. 2018. The state of world fisheries and aquaculture 2018 – Meeting the sustainable development goals. *CC BYNC-SA*, *3*.

FAO. 2024. *Fishery and Aquaculture Statistics –Yearbook 2021*, Rome, FAO Yearbook of Fishery and Aquaculture Statistics. doi: 10.4060/cc9523en

Idowu, S.O., Capaldi, N., Zu, L. and Gupta, A.D. Eds. 2013. *Encyclopedia of Corporate Social Responsibility*, Vol. 21, Berlin, Springer. doi: 10.1007/978-3-642-28036-8

Ito, M., Koba, K., Hikihara, R., Ishimaru, M., Shibata, T., Hatate, H. and Tanaka, R. 2018. Analysis of functional components and radical scavenging activity of 21 algae species collected from the Japanese coast, *Food Chemistry*, 255, 147–156. doi: 10.1016/j.foodchem.2018.02.070

Krause-Jensen, D., McGlathery, K., Rysgaard, S. and Christensen, P.B. 1996. Production within dense mats of the filamentous macroalga *Chaetomorpha linum* in relation to light and nutrient availability, *Marine Ecology Progress Series*, 134, 207–216. doi: 10.3354/meps134207

Lenzi, M., Persiano Leporatti, M., Gennaro, P. and Rubegni, F. 2017. Artificial top layer sediment resuspension to counteract *Chaetomorpha linum* (Muller) Kutz blooms in a eutrophic lagoon. Three years full-scale experience, *Journal of Aquaculture and Marine Biology*, 5(2), 00114.

Lenzi, M. 1992. Experiences for the management of Orbetello Lagoon: eutrophication and fishing. In *Marine Coastal Eutrophication*, pp. 1189–1198. Elsevier. doi: 10.1016/B978-0-444-89990-3.50102-4

Lenzi, M., Gennaro, P., Mercatali, I., Persia, E., Solari, D. and Porrello, S. 2013a. Physico-chemical and nutrient variable stratifications in the water column and in macroalgal thalli as a result of high biomass mats in a non-tidal shallow-water lagoon, *Marine Pollution Bulletin*, 75(1–2), 98–104. doi: 10.1016/j.marpolbul.2013.07.057

Lenzi, M. and Renzi, M. 2011. Effects of artificial disturbance on quantity and biochemical composition of organic matter in sediments of a coastal lagoon, *Knowledge and Management of Aquatic Ecosystems*, (402), 08.

Lenzi, M., Renzi, M., Nesti, U., Gennaro, P., Persia, E. and Porrello, S. 2013b. Vegetation cyclic shift in eutrophic lagoon. Assessment of dystrophic risk indices based on standing crop evaluations, *Estuarine, Coastal and Shelf Science*, *132*, 99–107. doi: 10.1016/j.ecss.2011.10.006

Lorentsen, S.H., Sjøtun, K. and Grémillet, D. 2010. Multi-trophic consequences of kelp harvest, *Biological Conservation*, 143(9), 2054–2062.

Lotze, H.K., Tittensor, D.P., Bryndum-Buchholz, A., Eddy, T.D., Cheung, W.W., Galbraith, E.D., Barange, M., Barrier, N., Bianchi, D., Blanchard, J.L. and Bopp, L. 2018. Ensemble projections of global ocean animal biomass with climate change, *BioRxiv*, 467175. doi: 10.1073/pnas.1900194116

Makkar, H.P. and Tran, G. 2016. Heuz e. V., Giger-Reverdin, S., Lessire, M., Lebas, F., & Ankers, P, pp. 1–17. doi: 10.1016/j.anifeedsci.2015.09.018

Marinho-Soriano, E., Fonseca, P.C., Carneiro, M.A.A. and Moreira, W.S.C. 2006. Seasonal variation in the chemical composition of two tropical seaweeds, *Bioresource Technology*, 97(18), 2402–2406.

Matos, G.S., Pereira, S.G., Genisheva, Z.A., Gomes, A.M., Teixeira, J.A. and Rocha, C.M. 2021. Advances in extraction methods to recover added-value compounds from seaweeds: sustainability and functionality, *Foods*, 10(3), 516.

McGlathery, K.J., Pedersen, M.F. and Borum, J. 1996. Changes in intracellular nitrogen pools and feedback controls on nitrogen uptake in *Chaetomorpha linum* (Chlorophyta) 1, *Journal of Phycology*, 32(3), 393–401. doi: 10.1111/j.0022-3646.1996.00393.x

Moreda-Pineiro, A., Pena-Vázquez, E. and Bermejo-Barrera, P. 2012. Significance of the presence of trace and ultratrace elements in seaweeds. In *Handbook of Marine Macroalgae*, pp. 116–170. doi: 10.1002/9781119977087

Olsson, J., Toth, G.B. and Albers, E. 2020. Biochemical composition of red, green and brown seaweeds on the Swedish west coast, *Journal of Applied Phycology*, 32(5), 3305–3317. doi: 10.1007/s10811-020-02145-w

Terasaki, M., Hirose, A., Narayan, B., Baba, Y., Kawagoe, C., Yasui, H., Saga, N., Hosokawa, M. and Miyashita, K. 2009. Evaluation of recoverable functional lipid components of several brown seaweeds (Phaeophyta) from Japan with special reference to fucoxanthin and fucosterol contents 1, *Journal of Phycology*, 45(4), 974–980. doi: 10.1111/j.1529-8817.2009.00706.x

van Hal, J.W., Huijgen, W.J.J. and López-Contreras, A.M. 2014. Opportunities and challenges for seaweed in the biobased economy, *Trends in Biotechnology*, 32(5), 231–233. doi: 10.1016/j.tibtech.2014.02.007

Wang, S., Zhao, S., Uzoejinwa, B.B., Zheng, A., Wang, Q., Huang, J. and Abomohra, A.E.F. 2020. A state-of-the-art review on dual-purpose seaweeds utilization for wastewater treatment and crude bio-oil production, *Energy Conversion and Management*, 222, 113253. doi: 10.1016/j.enconman.2020.113253

Ye, N.H., Zhang, X.W., Mao, Y.Z., Liang, C.W., Xu, D., Zou, J., Zhuang, Z.M. and Wang, Q.Y. 2011. 'Green tides' are overwhelming the coastline of our blue planet: taking the world's largest example, *Ecological Research*, 26, 477–485. doi: 10.1007/s11284-011-0821-8

6

CONTRIBUTION OF BIOTECHNOLOGY-BASED VALORISATION OF FORESTRY BY-PRODUCTS TO ACHIEVING SDG14

Dominic Duncan Mensah[a], Jeleel Opeyemi Agboola[b], Liv Torunn Mydland[a] and Margareth Øverland[a]

[a]Norwegian University of Life Sciences, Norway
[b]BioMar AS, Denmark

ABSTRACT

It is estimated that the largest share of future food fish will come from aquaculture production and that sustainable aquaculture is a precondition to realising this potential. Sustainable aquaculture will also play a key role in achieving several of the targets set out in SDG14. It is now established that most of the aquafeed ingredients used today are not sustainable and cannot support the projected growth of the sector, hence the need for sustainable alternatives. Sustainable aquaculture is multidimensional, therefore, this chapter focuses on sustainable feed ingredient sourcing. The authors explored a group of highly promising emerging novel ingredients known as microbial ingredients (MIs), means of producing them and how they can help achieve sustainable aquaculture and SDG14 targets. Specifically, the chapter narrows down on producing MIs from Norwegian spruce tree hydrolysates using a biotechnological approach and how Foods of Norway, a centre for research-based innovation at the Norwegian University of Life Sciences is leading efforts to produce feed-worthy MIs from industrial and agricultural by-products through biotechnology-based valorisation. MIs such as yeast, fungi, and bacterial meal can support the growth of Atlantic salmon without compromising the health of the fish. Thus, MI has a net positive impact on climate and can help achieve some targets in SDG14 by reducing pressure on marine resources used as fish feed ingredients. Suggestions on how to address current bottlenecks in scaling up MIs have also been provided in the chapter.

Keywords: Bio-refinery processing; forestry by-products; valorisation; microbial ingredients; salmonids; sustainability

INTRODUCTION

More than 600 million of the world's human population live in coastal areas while 2.4 billion people live within 100 km of the coastal zone for many reasons, including direct and indirect employment through recreation, trade, tourism, transportation, and fisheries. These marine-related activities are of great economic importance and contribute a high percentage of the gross domestic products (GDPs), especially in some developing countries. The oceans and seas are also sinks for more than half of anthropogenic carbon emissions, regulate the water cycle, and produce more than half the oxygen used by life on Earth. Unfortunately, the marine environment is under constant threat and overburdened due to the unsustainable rapid exploitation of these ecosystem resources and services. Just a little over five decades ago, many believed that marine fisheries were infinite. However, due to overfishing, illegal, unregulated and unreported (IUU) fishery-related activities, the percentage of biologically unsustainable fish stocks has increased from 10% in 1972 to 35.4% in 2019 (FAO, 2022). The marine environment faces several other challenges, including pollution, eutrophication through nutrient runoffs from farmland, and degradation of habitats and biodiversity, with accompanied economic losses estimated at some $83 billion per annum for the fishery sector and $6 billion for the aquaculture sector, mostly related to disease outbreaks (UNCTAD, 2018). With the current human population exceeding 8 billion people, the pressure on the marine environment is bound to increase. These problems are further aggravated by our changing climate.

In 2015, the United Nations (UN) and the global community adopted the sustainable development goals (SDGs) towards poverty eradication and increased prosperity among people by 2030. Of the 17 SDGs that were adopted, SDG14 mainly focused on the 'conservation and sustainable use of the oceans, seas, and marine resources thereof for sustainable development'. Ecosystem interactions, on one hand, are complex, which means a second axis of management response further heightens this complexity. One of the most important targets (target 14.4) of SDG14 is to enhance sustainable management of fisheries and ending harmful subsidies by 2020 to bring fish stocks to biologically sustainable levels by the year 2020 (UN, 2020). As of February 2020, whereas there are no reports as to whether the target was achieved, 70% of countries have ratified the agreement to join the efforts to combat IUU fishing as part of efforts to achieve this target. Even though achieving this target is extremely important, it must be admitted that wild capture fishery has plateaued, and that even if this target is achieved, production from capture fishery alone cannot bridge the demand gap driven by increasing human population and preference for fish protein. Aquaculture production continues to grow, reaching 87.5 million tonnes of live weight in 2020, accounting for 49.2% of total global fish production. Future food fish will largely come from sustainable aquaculture and will help achieve SDG14 by reducing pressure on wild capture fishery, reducing environmental footprint, and contributing to socioeconomic development.

Increased growth of aquaculture means a further increase in feed/feed ingredient supply, which has historically been met by fishmeal (FM) and fish oil (FO) obtained from low trophic wild capture fishery. Through the years, it has come to the fore that these ingredients cannot support the future growth of the aquaculture sector in a sustainable manner, thus there is a need for sustainable and responsible sourcing of novel feed ingredients. Therefore, this chapter addresses how biotechnology-based production of microbial ingredients (MIs) can help to increase the sustainability of salmonids aquaculture to help achieve several targets of SDG14. Also, this chapter includes initiatives taken by the Norwegian government, the Norwegian University of Life Sciences (NMBU), and how Foods of Norway, a research-based innovation centre at NMBU, is leading research on the use of sustainable ingredients in fish feeds.

EVOLUTION OF AQUAFEEDS

The early years of commercial aquaculture were marked with interests to increase the tonnage of production. However, as the sector continued to grow, interests widened to include environmental sustainability, animal welfare, product quality and the use of cost-effective feed ingredients. Aquafeeds have historically been reliant on FM and FO, which are considered 'gold standards' because of their high nutritional value. This class of feeds is referred to as 'Aquafeed 1.0'. FM is highly palatable with a favourable amino acid profile that meets the requirement of cultured fish. FOs, on the other hand, contain health-promoting long chain polyunsaturated omega-3 fatty acids such as docosahexaenoic acid (DHA) and eicosapentaenoic acid (EPA). During the past two decades, a decreasing supply trend of this class of aquafeeds have occurred (Glencross et al., 2023), even though they are still considered significant in the diets of carnivorous fish and some crustaceans. Despite their reduced supply, aquaculture is still considered the largest consumer of global FM and FO; 68% and 89%, respectively (Hua et al., 2019). With limited supply and increasing demand, the aquaculture industry evolved and gradually shifted to a new class of ingredients referred to as 'Aquafeed 2.0' to keep the supply of fish proteins to a growing global population.

Aquafeed 2.0 are predominantly land-based, comprising different proportions of grains and legumes (e.g. soybean meal and concentrate, gluten meal from corn and wheat, pea proteins, sunflower meal, rapeseed oil, etc.) as well as animal by-products such as chicken meal, bone, and meat meal. Between 1990 and 2020, the proportion of fish proteins in the feed for Atlantic salmon decreased significantly from 65.4% to 12.1%, with plant proteins as the main replacements (Aas et al., 2022a). During the last two decades, aquafeed 2.0 has been the main class of feed ingredients used in aquaculture because they are widely available at affordable prices. Plant proteins and their oils have several limitations in aquafeeds due to their relatively low protein content, low palatability, unbalanced amino acid profile, high-fibre content and the presence of a wide range of anti-nutritional factors (ANFs) that can affect growth performance and health of the fish. Life cycle assessment (LCA) studies showed that a large environmental footprint contributor in Atlantic salmon farming is related to feed transport and processing (Pelletier et al., 2009). In 2020, 1.98 million tonnes of feed ingredients were used in formulating Atlantic salmon diets in Norway (Aas et al., 2022a). Only 8% of these, mainly from FM and FO were obtained in Norway, while the remaining 92% were sourced from Europe and the global south, far away from where salmon farms are located. Although fish farming is categorised among the most climate-friendly animal husbandry, a carbon footprint of 7.9 kg CO_2 per kg of edible salmon product is still higher than chicken at 6.2 kg. Salmon farming will become more sustainable if the transport part of this environmental footprint is significantly reduced or eliminated. This implies that if aquaculture production is to help achieve SDG14, there should be less reliance on both Aquafeed 1.0 and 2.0. Achieving SDG14 through sustainable aquaculture will require technology-based innovation to source feed ingredients from unconventional means inorder to help address the challenges posed by the former classes of aquafeeds by a transition to an emerging new class of feed ingredient referred to as Aquafeed 3.0.

AQUAFEED 3.0: NORWEGIAN SUSTAINABILITY STORY

Norway is the largest producer of Atlantic salmon in the world, with about 7% annual growth since 1995, and contributed to about 2,700 million tonnes of salmon to the global fish supply in 2022 (Aas et al., 2022b). Globally, the salmon industry is the most industrialised and well-managed fish farming sector, serving as a standard for fish farming sectors in other countries. Atlantic salmon is a carnivore with a high protein and lipid requirement, which has traditionally been met by FM and FO. Currently, about 30% of FM used in aquaculture is obtained from by-products of fishery and aquaculture that are not suitable for human consumption (Aas et al., 2022a), hence reducing the pressure on capture fisheries, making these resources more sustainable. Over the last years, the Norwegian salmon industry has been adjudged the most sustainable protein producer by the 'coller FAIRR index'. Despite this feat, concerted efforts are taken by the government and the industry to increase the sustainability of the sector to meet the ever-growing public demand for sustainability in salmonids production.

Among the strategies is to reduce food waste by 50% by 2030, since if less food goes to waste, then less will have to be produced on land and from the ocean and a resultant reduction in GHG from the production or importation. Perhaps the most important part of this action plan in this context is the government mandate that all feed ingredients should come from sustainable sources by 2030. Sustainable sources imply the use of locally available Norwegian resources, especially by upcycling organic waste streams and waste gasses into these novel ingredients to offset carbon in the atmosphere. Within this same period, the national ambition is to increase the current production of Atlantic salmon from 2 to 5 million tonnes by 2050, which will require 6.2 million tonnes of feed ingredients (Aas et al., 2022b). However, several barriers exist, including the availability of the local resources to produce enough ingredients to meet the needs of the aquaculture industry and the fact that developing novel feed ingredients takes time. This challenge creates an opportunity to explore the Aquafeed 3.0 class and novel feed ingredients. To meet this 2030 mandate, there is need for a close collaboration between the academia, industry, the political establishment, regulatory authorities, and every stakeholder along the whole value chain to improve the sustainability of aquaculture and to achieve SDG14.

Aquafeed 3.0 consist of a class of novel feed ingredients that includes microbial proteins (previously known as single cell proteins, SCP), and microbial oils (previously known as single cell oils, SCO), insect meals, fermented by-products, seaweed as well as fishery and aquaculture processing by-products. MIs and novel aquafeeds can help to improve the sustainability of aquaculture since their production is about converting low value resources to high-value ingredients using a biotechnological approach. The production of novel feed ingredients falls in the circular bioeconomic framework (Fig. 1), within which renewable biological products are used to manage the food, land, and industry by keeping the value of biomass in a cascading groove. It is estimated that 1.3 billion tonnes of food waste is generated per annum globally, which is almost a third of all food produced (FAO, 2021). These waste food products cannot be fed directly to fish, but biotechnology-based valorisation, bioconversion, and biotransformation of these resources into high-value MIs can provide high-quality nutrients for fish growth and development demonstrated in later sections of this chapter. They can also be used as substrates for insects such as black soldier fly larvae, which can be incorporated into aquafeeds. Thus far, some of these alternatives including fish processing by-products as well as SCO are already used in commercial fish diets. However, the full potential of these alternatives is not yet fully realised. Currently, only 0.4% of such novel feed ingredients (8,130 tonnes of total feed ingredients) were included in the Norwegian salmon diets in 2020 (Aas et al., 2022b). This implies that a lot of effort between now and 2030 is needed to ensure cost-effective production and to remove bottlenecks to ensure scalability of these ingredients if the Norwegian government mandate is to be realised. Over the last decade, the NMBU and the research team at the Centre of research-based innovation 'Foods of Norway' have studied the suitability of some of these novel feed ingredients, which have shown promising results in salmonids, as well as for land-based farm animals. Among several ingredients evaluated, this chapter will focus solely on novel MIs including yeast, bacteria, and fungi.

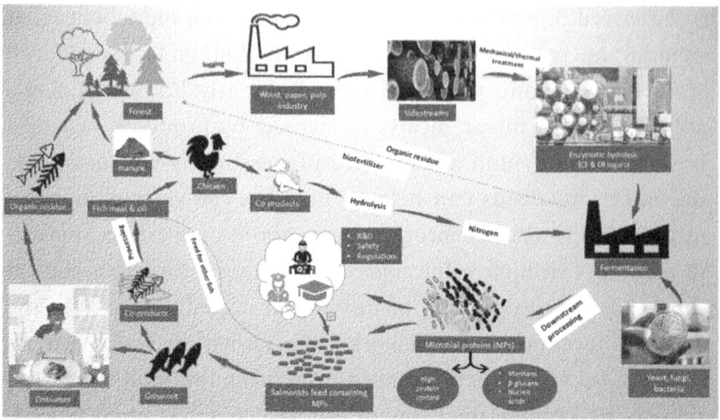

Fig. 1. Conceptualising the Use of Spruce Wood Hydrolysates as Substrates for Producing Microbial Proteins Through a Circular Bioeconomic Framework and How They Can Help to Achieve SDG14.

BIOTECHNOLOGY-BASED PRODUCTION OF MIs

Only 3% of the total land mass of Norway is farmland, while 39% is covered by forest, dominated by coniferous plants such as spruce tree, *Picea* sp. The main reason for the sustainable harvesting of spruce trees in Norway is for construction. About a third of the wood results in sawdust and chippings in sawmills. With the advancement of technology over the years, utilisation of such tree side-streams including those of paper pulping is increasing. For example, Borregaard, a Norwegian forest biorefinery company using up to 92% of the wood by-products to produce biopolymers, biovanillin, cellulose, and bioethanol used in the food, cosmetics, and the energy sectors. The side-streams produced from these industrial processes contain 5- and 6-carbon sugars which have myriad applications in agriculture, pharmaceuticals, foodstuffs, and aquaculture. Second-generation sugars produced from these processes can be used as a substrate for producing yeast, bacteria, fungi, and microalgae. The review of Øverland and Skrede (2017) outlined three major steps through which second-generation sugars can be obtained from lignocellulosic biomass such as spruce trees or agricultural by-products. The conversion of lignocellulosic biomass into sugars starts with pre-treatment using thermo-chemical methods to break down hemicellulose-lignin complexes to individual hemicellulose and lignin. This is followed by enzymatic hydrolysis of these complex compounds to C5 and C6 monosaccharides. The sugars are then used as substrates in the fermentation of yeast or filamentous fungi, followed by several down-stream processes to obtain high-quality microbial proteins. The crude protein content of yeast ranges between 38% and 60%, while that of filamentous fungi such as *Paecilomyces variotii* could be as high as 65%. The protein content of *P. variotii* is like highly refined soy protein concentrate (SPC) which is widely used in diet of Atlantic salmon. The MIs have favourable amino acid composition that can meet the requirements of salmonid except sulphur containing amino acids such as methionine and cysteine, which may be supplemented when MIs are used as replacement for FM in salmon diets. It is noteworthy that amino acid supplementation is not new but a common practice when plant proteins are used in fish diets.

MIs also have a high content (10–15%) of non-nitrogen protein, such as nucleic acids, which might have protein-sparing effect (Agboola et al., 2021). Additional advantages of MIs are their bioactive components such as mannans, β-glucans, as well as nucleic acids, which are documented immunostimulants and prebiotics. Therefore, these MIs also have additional properties beyond their nutritional values, such as prophylactics. These prophylactic properties of MIs can serve as alternatives to antibiotics and chemotherapeutants to treat fish related diseases.

MICROBIAL PROTEINS CAN SUSTAIN THE GROWTH OF SALMONIDS

According to Glencross et al. (2007), a feed is only as good as its ingredients. In the same vein, MIs are only as good as the substrates on which they are fermented and the down-stream processing they go through. The nutritional content of MI is reflective of the composition of the substrate on which they are grown, which makes it easy to manipulate these ingredients to fit the nutritional profile of interest. This property of MI as well as their health-promoting effects gives them a market competitive advantage over other novel feed ingredients. Nutrient characterisation, digestibility, palatability, utilisation, and health-added values are important parameters when considering the suitability of ingredients as aquafeed resources. These feed ingredient properties have been documented for several yeast and bacteria species and strains. Overall, a net positive effect on growth performance has been reported in salmonids-fed diets partially replaced with *Saccharomyces cerevisiae* (Hansen et al., 2021), although digestibility was lower compared with the FM control diets. Earlier studies have shown that non-*saccharomyces* yeast species including *Cyberlindnera jadinii* and *Kluyveromyces marxianus* supported the growth of Atlantic salmon (Øverland et al., 2013). These yeast species resulted in better nitrogen retention and digestibility like those of the FM control diets, proven to be suitable alternative feed ingredients in the diets of these valuable fish species.

A high nitrogen (N) digestibility and utilisation are important for the health of aquatic ecosystems. Poorly digested and utilised N in feed results in excess N load in the culture environment of fish and may lead to algal blooms which may be toxic, lower the species diversity beneath fish culture cages or lead to hypoxic conditions. Therefore, yeast species such as *C. jadinii* and *K. marxianus* with improved digestibility are more appealing as viable salmonid feed ingredients. Yeasts have recalcitrant cell walls matrix with several polymers which are poorly digestible and bind the proteins in a matrix which prevent access of digestive enzymes, a plausible reason for the poor protein digestibility reported for *S. cerevisiae*. In later studies by Hansen et al. (2021), different down-stream processes were employed to break down the cell wall of yeast to improve protein digestibility.

Another emerging and highly promising microbial protein ingredient is filamentous fungi *P. variotii* (PEKILO®) which was first used in the 60s to the 90s as a means of dealing with spent sulphite liquor from the pulping industry in Finland. What makes *P. variotii* more attractive as a potential feed ingredient is its high crude protein content (55–64%), satisfactory growth in culture conditions, easy separation from the liquor and no reported toxicity in test animals, compared with other microfungi. It can grow on a range of industrial waste streams such as molasses, wood hydrolysates, spent sulphite liquor and vinasse. Aside the high protein content, the cell wall contains up to 24% insoluble carbohydrates of which 14% are β-glucans with potential health beneficial effects. *P. variotii* has been used in the diets for growing and finishing pigs, chicken, and lactating calves. Foods of Norway are currently evaluating the nutritional value and health effects of *P. variotii* in Atlantic salmon.

Microbial proteins are not limited to fungi and yeast, but also include bacteria. The technology to produce bacteria meal from waste was available as early as the 70s, with notable commercial bacteria meal including Pruteen®, produced from *Methylophilus methylotrophus* through methanol oxidation. Interestingly, this product has a high protein content up to 72% with amino acid composition comparable to that of FM and a high nucleic acid content (Braude et al., 1977). No difference in growth rate, N retention, apparent digestibility or amino acid compositional difference between pigs fed Pruteen® and FM-supplemented diets was observed. A 2010 review article by Øverland et al. (2010), showed that another bacteria, *Methylococcus capsulatus*, grown on natural gas (methane) as a substrate could serve as a protein source for several fish species including Atlantic salmon, rainbow trout and Atlantic halibut as well as livestock (broiler chicken, pigs, mink). As in yeast and fungi, bacteria meals could contain immunostimulants such as nucleic acids (Mydland et al., 2008), lipopolysaccharides, peptidoglycans, flagellin from flagellated bacteria, and β-glucans that are beneficial to fish health including counteracting soybean meal-induced enteritis (Romarheim et al., 2013). A healthier fish is more efficient at converting feed into biomass, reduces nutrient losses to the environment and lowers the amount of input cost due to these merit of MIs over other novel ingredients.

HOW MICROBIAL PROTEINS CAN HELP ACHIEVE SDG14

An advantage of some MI such as microalgae and photosynthetic bacteria is that they could have climate mitigation effects due to their ability to convert waste gasses (biogenic CO_2, CO, and CH_3) into useful proteins and lipids, implying a net positive effect on the environment through absorbing greenhouse gases from the atmosphere. The global warming potential of methane is 27–30 times higher than carbon dioxide over a 100-year span. For that matter, producing bacteria meal such as *M. capsulatus* from methane as opposed to their use for heating is an efficient way of reducing the concentration of methane or CO_2 from its combustion in the atmosphere and averting any climate damage it would have caused. MIs also provide alternative means of converting side-streams and nutrient-rich effluents from forestry, animal husbandry and agriculture into high-quality proteins that otherwise would pollute the marine environment. For instance, spent sulphite liquors from the paper pulping industry usually contain sugars from hemicellulose and xylose. Other compounds that could be present include furans, organic acids, and inorganic residues. Their lignocellulosic co-products may contain lignosulfonates, polyhydroxybutyrates, ethanol and xylitol. Improper disposal of these effluents can lead to pollution of water bodies and ultimately the marine environments, while MIs provide a solution to this problem. In addition, MI can be produced all year around since their production is climate independent, requires no arable lands, and uses little water. For this reason, the supply of microbial proteins for fish feed can be constant, unlike FM, FO, or plant proteins whose prices are highly volatile due to fluctuations in their supply caused by geopolitics or climate-driven factors. A major criticism of aquaculture is the direct competition between feed and food for direct human consumption. For example, small pelagic fish such as sardines, anchovies, and herrings are the largest sources of FM and FO. Therefore, microbial proteins as alternatives will reduce the pressure on the wild stock, institute a balance in marine food chains and help achieve SDG14. Because MI can be produced on a range of local bioresources in proximity to fish farms or feed processing plants the environmental impact due to transport is minimised. Another major issue with the use of plant proteins in salmonid feeds is that their production requires the use of land, water, and input fertilisers. Currently, SPC from soy which are largely imported from Brazil, makes up to 20% of salmonid diets. High dependency on soybeans will lead to more deforestation, less carbon sequestration, and increase in global warming, which affects everything including the oceans. Again, runoffs from farmlands could contain N and phosphorus which causes eutrophication, algal blooms, and dead zones in the marine environment. All these potential problems can be eliminated if MIs are well-developed and used as replacements of both Aquafeed 1.0 and 2.0. Aquaculture production should be climate-friendly, increase food production while at the same time ensuring sustainability and this can be achieved with the use of microbial and novel feed ingredients.

EXPERIENCE AS A MASTER'S STUDENT AT NMBU

At NMBU, study programs and course curricula are designed to address the UN SDGs including SDG14 at the bachelor, master, and doctorate levels. This means almost all students are aware of the UN SDGs and the underlying problems. Students are trained with insights and knowledge on the relationship between effective production, economic yield, and the need for cost-effective and sustainable feed ingredients. Through the feed technology course, students gain both theoretical and practical knowledge (using the feed processing plant on campus) on various processes involved in aquafeed production. Molecular courses related to fish metabolism and health assessment are designed to engage the interests of students and to widen their scope on ingredient evaluation beyond nutrition. The Foods of Norway group has given many students (both local and international) the opportunity to obtain hands-on experience on designing fish experiments on nutritional and health assessments through a master thesis, most of which involve the evaluation of MIs along the value chain from processing until the final fish product. Like many other students, the author is a proud product of this robust system of producing and evaluating MI whose master thesis was based on evaluating the health effects of two yeast species *C. jadinii* and *W. anomalus* subjected to different down-stream processing on Atlantic salmon. This is also the basis for his current position as a PhD student, testing the health benefits of a filamentous fungi produced using this biotechnology-based approach to the health of Atlantic salmon.

The Foods of Norway research team has done much in raising awareness of the need to find suitable replacements through podcasts, seminars, and media coverage. A large proportion of the Norwegian public are aware of the salmon feed ingredient challenges and efforts that are being put in place to address these challenges thanks to this awareness creation. Consumer acceptance is very necessary. Through awareness creation and transparency, public trust will be enhanced should MI come into full effect. Also, collaboration between research groups at NMBU and with the industry is vibrant and effective, which is extremely important, especially if large-scale production of MI is going to take effect. These industries will advance large-scale production, making their early involvement integral.

CONCLUSION

The use of MI as a sustainable feed ingredient in salmonids feeds can have a significant positive effect on achieving several of the targets of SDG14. However, a few challenges exist that need to be addressed to realise this opportunity. Many waste streams that could serve as feedstock to produce these MIs are now restricted by EU regulations. This is not exclusive to MI, but also other green proteins such as insects. Therefore, continued research on the safety evaluation of alternative substrates to produce these microbes to meet volume demands for the salmonids aquaculture industry is needed.

Other substrates such as waste gasses (e.g. hydrogen and carbon dioxide) can also be feasible substrates for the fermentation of microbes. However, this fermentation technology needs to be further developed. Thus, the fermentation technology needs to be improved to efficiently utilise a range of alternative feedstock. Competition for the substrate for non-feed uses such as fuel is another area that demands regulatory control. For example, competition of biomass and waste streams as input factors for bioenergy or as feedstock to produce yeast, fungi or bacteria will strongly affect the potential to scale up MIs for future fish production.

This competition is bound to increase, owing to increased human population and climate changes. Therefore, these regulations need to make these accessible for the course of producing MIs to help realise SDG14.

REFERENCES

Aas, T.S., Åsgård, T. and Ytrestøyl, T. 2022a. Utilization of feed resources in the production of Atlantic salmon (*Salmo salar*) in Norway: an update for 2020, *Aquaculture Reports*,26, 101316.

Aas, T.S., Ytrestøyl, T. and Åsgård, T.E. 2022b. Utnyttelse av fôrressurser i norsk oppdrett av laks og regnbueørret i 2020. Faglig sluttrapport, *Nofima rapportserie*.

Agboola, J.O., Øverland, M., Skrede, A. and Hansen, J.Ø. 2021. Yeast as major protein-rich ingredient in aquafeeds: a review of the implications for aquaculture production, *Reviews in Aquaculture*,13(2), 949–970.

Braude, R., Hosking, Z., Mitchell, K., Plonka, S. and Sambrook, I. 1977. Pruteen, a new source of protein for growing pigs. I. Metabolic experiment: utilization of nitrogen, *Livestock Production Science*,4(1), 79–89.

FAO.2021.*Food loss and food waste*, Rome, FAO.

FAO. 2022. *State of World Fisheries and Aquaculture. Towards Blue Transformation*, FAO, Rome.

Glencross, B., Fracalossi, D.M., Hua, K., Izquierdo, M., Ma, K., Øverland, M., Robb, D., Roubach, R., Schrama, J. and Small, B. 2023. Harvesting the benefits of nutritional research to address global challenges in the 21st century, *Journal of the World Aquaculture Society*.

Glencross, B.D., Booth, M. and Allan, G.L. 2007. A feed is only as good as its ingredients – A review of ingredient evaluation strategies for aquaculture feeds, *Aquaculture Nutrition*, 13(1), 17–34.

Hansen, J.Ø., Lagos, L., Lei, P., Reveco-Urzua, F.E., Morales-Lange, B., Hansen, L.D., Schiavone, M., Mydland, L.T., Arntzen, M.Ø. and Mercado, L. 2021. Down-stream processing of baker's yeast (*Saccharomyces cerevisiae*) – effect on nutrient digestibility and immune response in Atlantic salmon (*Salmo salar*), *Aquaculture*,530, 735707.

Hua, K., Cobcroft, J.M., Cole, A., Condon, K., Jerry, D.R., Mangott, A., Praeger, C., Vucko, M.J., Zeng, C. and Zenger, K. 2019. The future of aquatic protein: implications for protein sources in aquaculture diets, *One Earth*, 1(3), 316–329.

Mydland, L., Frøyland, J. and Skrede, A. 2008. Composition of individual nucleobases in diets containing different products from bacterial biomass grown on natural gas, and digestibility in mink (Mustela vison), *Journal of Animal Physiology and Animal Nutrition*,92(1), 1–8.

Øverland, M., Karlsson, A., Mydland, L.T., Romarheim, O.H. and Skrede, A. 2013. Evaluation of *Candida utilis*, *Kluyveromyces marxianus* and *Saccharomyces cerevisiae* yeasts as protein sources in diets for Atlantic salmon (*Salmo salar*), *Aquaculture*,402, 1–7.

Øverland, M. and Skrede, A. 2017. Yeast derived from lignocellulosic biomass as a sustainable feed resource for use in aquaculture, *Journal of the Science of Food and Agriculture*,97(3), 733–742.

Øverland, M., Tauson, A.-H., Shearer, K. and Skrede, A. 2010. Evaluation of methane-utilising bacteria products as feed ingredients for monogastric animals, *Archives of Animal Nutrition*,64(3), 171–189.

Pelletier, N., Tyedmers, P., Sonesson, U., Scholz, A., Ziegler, F., Flysjo, A., Kruse, S., Cancino, B. and Silverman, H. 2009. *Not All Salmon Are Created Equal: Life Cycle Assessment (LCA) of Global Salmon Farming Systems*. ACS Publications.

Romarheim, O.H., Hetland, D.L., Skrede, A., Øverland, M., Mydland, L.T. and Landsverk, T. 2013. Prevention of soya-induced enteritis in Atlantic salmon (*Salmo salar*) by bacteria grown on natural gas is dose dependent and related to epithelial MHC II reactivity and CD8α+ intraepithelial lymphocytes, *British Journal of Nutrition*,109(6), 1062–1070.

UN. 2020. *Progress towards the Sustainable Development Goals* E/2020/57).

UNCTAD. 2018. *Achieving the Targets of Sustainable Development Goal 14: Sustainable Fish and Seafood Value Chains and Trade*, Geneva, United Nations.

7

MARINE PROTECTED AREAS, COASTAL AND MARINE MANAGEMENT

Chris de Blok[a] and Richard Page[b]

[a]*MatureDevelopment bv. Netherlands*
[b]*Palau Community College, Palau*

ABSTRACT

Sustainable Development Goal 14 of the United Nations aims to 'conserve and sustainably use the oceans, seas and marine resources for sustainable development'. To achieve this goal, we must rebuild the marine life-support systems that provide society with the many advantages of a healthy ocean. Therefore, countries worldwide have been using Marine Protected Areas (MPAs) to restore, create, or protect habitats and ecosystems. Palau was one of the first countries to use MPAs as a tool to develop biodiversity within its exclusive economic zone. On 22 October 2015, Palau placed approximately 80% of its maritime territory in a network of locally monitored MPAs, which has now shown a population increase in stationary and migratory fish species. This movement towards a MPA was intentional and because of increased pressure from tourism and the increasing incursion of foreign fishing vessels in Palauan territorial waters. Since countries worldwide are using and looking towards

MPAs, secondary protection projects are becoming more and more popular. This chapter highlights the practical implementations and results in Palau, how to theoretically apply this within the Greater North Sea in combination with Windmill Farms, and how the Marine Strategy Framework Directive stimulates these practices.

Keywords: Marine ecology; marine protected areas; windmill farms; protected area networks; Palau National Marine Sanctuary; Greater North Sea; Marine Strategy Framework Directive; spill-over effect

INTRODUCTION

The United Nations' Sustainable Development Goal 14 aims to 'conserve and sustainably use the oceans, seas, and marine resources for sustainable development'. To accomplish this, we must reconstruct the mechanisms that sustain marine life and give society the numerous benefits of a healthy ocean. In order to repair, create, or safeguard habitats and ecosystems, countries worldwide have adopted marine protected areas (MPAs) (Duarte et al., 2020). Palau, as displayed in Fig. 1, was one of the pioneering nations to employ MPAs to foster biodiversity within its exclusive economic zone (EEZ). On 22 October 2015, Palau designated a locally supervised MPAs network to cover about 80% of its marine territory (Cimino et al., 2019; Jaiteh et al., 2021). Beginning that year, each state in Palau started enlisting local populations in MPAs, teaching them in such areas, and designating exclusive coastal areas where traditional and contemporary fishing techniques were rigorously restricted. Due to growing pressure from tourism and the intrusion of foreign fishing vessels into Palauan territorial seas, there was an intentional drive towards establishing a MPA. The MPAs stated political goal was to increase and enrich the biodiversity of the nations-controlled territorial waters.

MPAs are a conservation benefit to fish assemblages, including no-take zones, limited fishery access, and protection of traditional and indigenous fishing practices. The development of MPAs also benefits neighbouring areas through 'spill-over' (Di Lorenzo

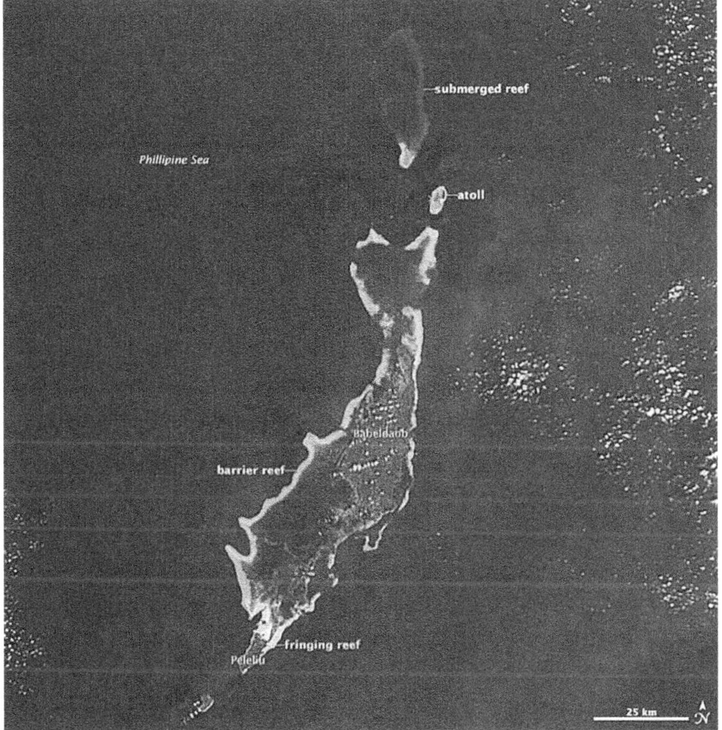

Fig. 1. An Aerial Satellite Picture of Palau (Schmaltz and Voiland, 2012).

et al., 2016). Palau's MPAs created a dramatic spill-over effect in the coral triangle of Micronesia, an area stretching from Guam to Yap and Palau. This has made Palau one of the countries with the highest numbers of pelagic species that pass through Palau's EEZ. These species include all species of Pacific tuna, mackerel, shark, barracuda, snapper, bream, marlin, and groupers, as well as reef fish such as rabbitfish and surgeon fish.

Marine spatial planning creates multipurpose areas that do not have to be out at sea and can also be used along the coastal areas. The Netherlands can use the dikes as biological hotspots that can use the same effects in these ways – working towards achieving the targets 14.2, 14.4, 14.5, 14.6, 14.a, and 14.b. In this chapter, the authors aim to inspire the young generation, professionals, and decision-makers to create system-based approaches to marine ecology and human economy that can sustainably increase biodiversity while allowing traditional cultural and environmental systems to remain intertwined.

POSSIBILITIES FOR ECOSYSTEM-BASED MANAGEMENT IN THE GREATER NORTH SEA

In the book, *The Deep*, Alex Rogers describes bottom trawling as a highly destructive method of fishing, with some methods even capable of destroying the volcanic rock of certain seamounts (Rogers, 2019). Especially within the Greater North Sea (Fig. 1), this type of trawling is the most practised form of fishery. Consequently, a loss of natural hard substrates is described by Prof Dr Han Lindeboom (Lindeboom, 2015) and an enormous shortage of living space for most benthic animals was observed. These animals form the basis of the food chain and are therefore essential for the productivity of our oceans.

An understanding of the current spatial planning is needed to assess the impact of MPAs fully. In total, according to the OSPAR (2018) status report, a total of 142,589 km^2 (18.6%) of the Greater North Sea falls under the Marine Protected Status (OSPAR, 2018). Apart from MPAs, there are also areas for planned or active Offshore Windmill Farms (OWFs) in the Greater North Sea (Fig. 2). In total, 41 active windfarms are presented in the Greater North Sea; however, in Fig. 3 there are large development areas for potential windfarms (Chirosca et al., 2022). The location and distribution of these windmills play an important role in the possibility of using OWFs for MPAs and potentially spill-over, this is because spill-over in the European seas has only been detected close to OWFs (Weigel et al., 2014; Halouani et al., 2020).

European Stimulation Through the Marine Strategy Framework Directive

The drive towards sustainable solutions continue to grow. Ashley et al. (2014) note that by 2024 the rapid expansion of offshore wind farming will reach a target set by the European Parliament (The European Parliament; Council of the European Union, 2009)

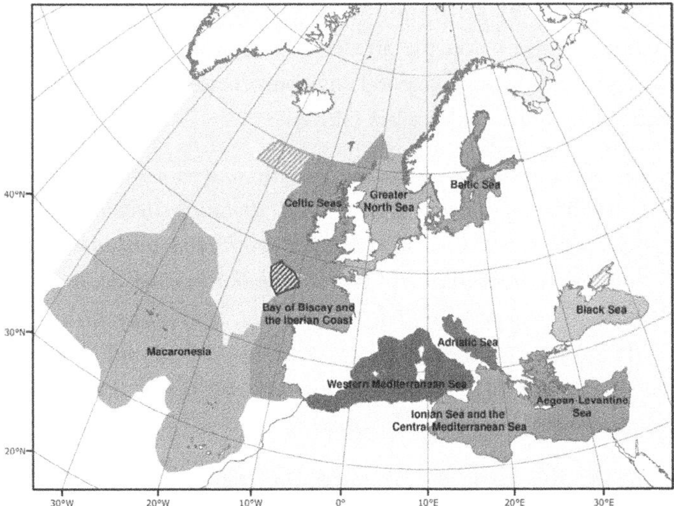

Fig. 2. Representation of the Marine Regions and Subregions of the Marine Strategy Framework Directive (MSFD) Article 4 (Jensen and Panagiotidis, 2018).

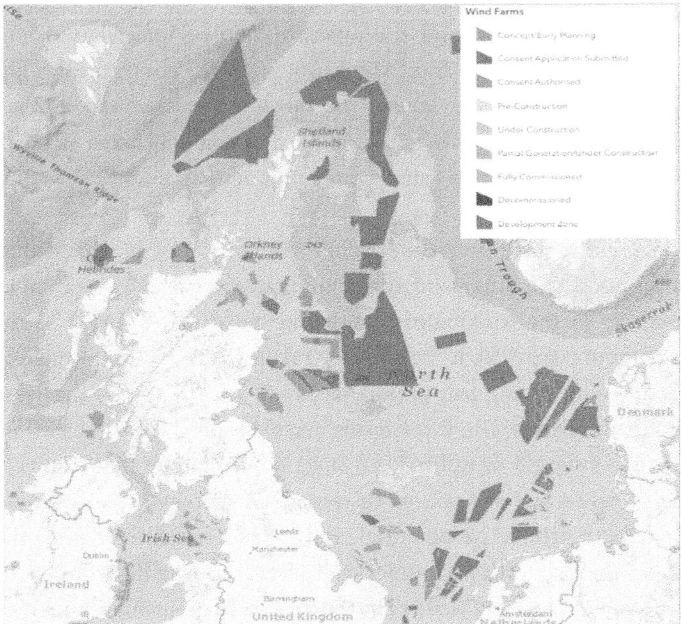

Fig. 3. A Map of All Developing, Under Construction, and Constructed OWFs in the North Sea (Chirosca et al., 2022).

of 20% of energy generation from renewables by 2020. There is also the European Marine Strategy Framework Directive, whose primary goal is to achieve good environmental status (GES) for EU waters. The directive describes GES as

> *the environmental status of marine waters where these provide ecologically diverse and dynamic oceans and seas which are clean, healthy and productive within their intrinsic conditions, and the use of the marine environment is at a level that is sustainable, thus safeguarding the potential for uses and activities by current and future generations. (The European Parliament; Council of the European Union, 2008, Article 3)*

To achieve this goal, several descriptors were identified of which five are relevant to the use of OWFs for spill-over.

Firstly, there is 'Biodiversity', which refers to the diversity of life on Earth. This diversity can be seen at many scales, including genetic variety, species, and ecosystems. Therefore, the term refers to all living things, their behaviours, the surroundings or habitats in which they exist, and the intricate web of interrelationships among them, such as food webs and resource competition (The European Parliament; Council of the European Union, 2023a). This descriptor can only be accomplished when all other descriptor goals have been met since almost all activity could have some effect on biodiversity.

Secondly, there are economically exploited stocks which must be in good condition and that exploitation must be sustainable, resulting in the Maximum Sustainable Yield (MSY). MSY is the maximum yearly catch that may be taken without diminishing the productivity of the fish population year after year. On the website of the Descriptor, three terms are mentioned to achieve MSY, (1) stocks should be utilised in a sustainable manner that results in high long-term yields, (2) to sustain stock biomass, stocks should have full reproductive potential, and (3) as a measure of a healthy stock, the proportion of older and larger fish/shellfish should be maintained (or increased) (The European Parliament; Council of the European Union, 2023c).

Furthermore, the relationships between species in a food web are complicated and continually changing, identifying a single state that reflects a GES is challenging. Moreover, because the food web is a fully integrated system, stresses on one area may have unanticipated consequences elsewhere (Rogers et al., 2010). For example, by harvesting shellfish like mussels or oysters, which are habitat-building keystone species, a collapse of habitat creation or prey for predators could influence entire fish stocks (Dame and Kenneth, 2011; Wijsman et al., 2018; Evers, 2022). Therefore, there could potentially be unintended implications for various other species. Managing human activity to achieve the required balance of species in the system is thus a significant task.

The descriptors also mention that the seafloor integrity should be protected because the seafloor is an important compartment and platform for marine life in the ocean. After all, it has a high biomass productivity, especially in shallow seas. This seafloor creates several habitats for permanent or mobile marine organisms that dwell inside and above the sea floor. Maintaining seafloor integrity is thus essential for preserving marine biodiversity and living resources. According to the task group 6 Report of J. Rice, 'Sea Floor includes both the benthic community's physical structure and biotic composition. Integrity includes the characteristic functioning of natural ecosystem processes and spatial connectedness' (Rice et al., 2010). They indicate that it is essential to create a good integrity balance between abiotic and biotic factors. To characterise the seafloor, distinct seabed types are commonly distinguished in the directive based on depth, substrate type, and species composition (The European Parliament; Council of the European Union, 2023d).

Lastly, the 'Introduction of energy, including underwater noise, should be at levels that do not adversely affect the marine environment' (The European Parliament; Council of the European Union, 2023b). Underwater the light does not travel far. Therefore, marine fauna is dependent on sound. The marine fauna perceive their world through sound pressure and vibrations. This makes it so that the impact of natural sound, being a wide range of marine fauna, waves, rain, wind, and seabed movement, and anthropogenic sounds from human activity at sea include shipping and other marine craft, building and installations, sonar and seismic surveys is the very thing that forms the perception of marine fauna from the marine environment (Flanders Marine Institute [VLIZ], 2020; Thomsen et al., 2021).

Marine Protected Areas and Offshore Wind Parks

As described the Greater North Sea has a lot of MPAs and existing or planned OWFs. Overall MPAs are implemented worldwide as an efficient tool to preserve biodiversity and protect ecosystems (Colléter et al., 2014; Soukissian et al., 2021). Furthermore, different forms of policy are used in different MPAs, for example, there are MPAs that have a completely no-take policy, and there are MPAs that have regulated policies so that activity is allowed (Sweeting and Polunin, 2005). According to Benjamin S. Halpern (2003), marine reserves of any size improve density, bio-mass, individual size, and variety in all functional categories. He describes that the density of organisms is doubled, while biomass is nearly tripled in reserves.

Besides improving the biological conditions within the MPAs, they could also improve conditions outside. This spill-over effect can benefit fisheries outside of the marine reserves. However, spill-over and MPAs are not a cure-all for fisheries management (Sweeting and Polunin, 2005). It has an important role within management. Although this sounds like the best solution for fisheries management as far as the spill-over effect is known, it only works directly outside of the MPA and is therefore highly localised (Halouani et al., 2020). Therefore, closing off one area is ineffective, however, a network of MPAs could theoretically create a beneficial effect (Weigel et al., 2014). This is where OWFs come into effect.

To create a collaborative network in the Greater North Sea some flexible thinking and trade-offs are needed. Therefore, OWFs can function as an MPA (Ashley et al., 2014; Hammar et al., 2016). They can help build this collaborative network through this function and create a situation limiting the number of fishing grounds taken if the need for green energy and habitat protection are not combined.

THE MPA APPROACH OF PALAU

As early as 2003, the legislature of Palau began the process of designation of MPAs through the structure of Protected Area Networks (PANs). This process had the express intent of developing a system of conservation areas in Palau that could be monitored by a statewide network of protected area offices and advisors under the direction of, PAN and the State Governments (Gouezo et al., 2016). This process was developed in response to a public perceived loss of biodiversity and heightened tourism flows, primarily from larger states in Asia (Medel, 2020). This followed a series of legislative actions supported by traditional leaders and the Palauan community codified in RPPL 6-39 and subsequently expanded in the Palauan code in RPPL 7-42 to establish funding for the PAN and begin the legislative work of developing a larger MPA. The use of MPAs as an effective conservation tool was the law's express intent. Creating these areas as intentional acts towards environmental conservation allowed Palauans to take a leading role in conserving their resources threatened by external or internal risk of over-exploitation (Fujita et al., 2022).

It is feasible that the area between Guam, Yap, and Palauan territorial waters today, historically acted as the nursery for many fish and sea cucumber species in the western central Pacific region. From this aspect, conservation has been recognised as part of the traditional cultural practices in these islands for many centuries. The encounter with outside systems, colonisation, World War II, trusteeship, and the increasingly globalised economy of Palau necessitated a legal framework to encode traditional conservation and the use of protected areas was viewed to manage fish stocks in the face of largely unregulated commercial fishing. Assessments of biological diversity were needed which scientists at the Micronesian Mariculture Demonstration Center (MMDC), beginning in 1976, had suggested since 1987 (MMDC *The Prompt Report of the Fifth Scientific Survey of the South Pacific*). With the deficit in existing legal and protective frameworks and commensurate Palauan Independence in 1994, the movement towards environmental law frameworks to protect marine resources was strong (Ueki, 2000).

The law catalysed a broader legislative push to protect as much of Palau's EEZ as possible. At the time, the movement towards creating a National Marine Reserve was being formulated by Noah Idechong and the Palau Conservation Society working with the Pew Charitable Trust, Palauan diaspora, and the United States. The use of MPAs was further expanded in 2007 with the formulation of the Micronesia Challenge, and in 2008 by the 7th Congress of Palau with the establishment of the PAN fund (PANF). This was rapidly followed in 2010 by implementing a 'green fee' system that assesses tourists to the island with a fee for the use of the environment. Between early 2011 and March 2012, the PANF and PAN system elected a board of directors. It began an effective mobilisation of resources directed towards the designation, management, and allocation of funding for protected areas.

The series of commitments in SDG 14 and the Micronesia Challenge led to a commitment by the Republic of Palau to expand its MPA network further to effectively cover at least 30% of its coastal resources by 2020. On 22 October 2015, Palau, with President Tommy Remengasau's leadership, initiated a much broader plan that had been building since 2008 with the creation of one of the largest marine sanctuaries in the world. The Palau National Marine Sanctuary (PNMS) effectively placed 80% of Palau's EEZ or 500,238 km sq under protection, while reserving an unaffected coastal area of 18,628 km sq for indigenous and traditional reef fishing practices and designation of 85,896 km sq zone for domestic pelagic fishing activity (Figs. 4 and 5) (Cimino et al., 2019; Jaiteh et al., 2021).

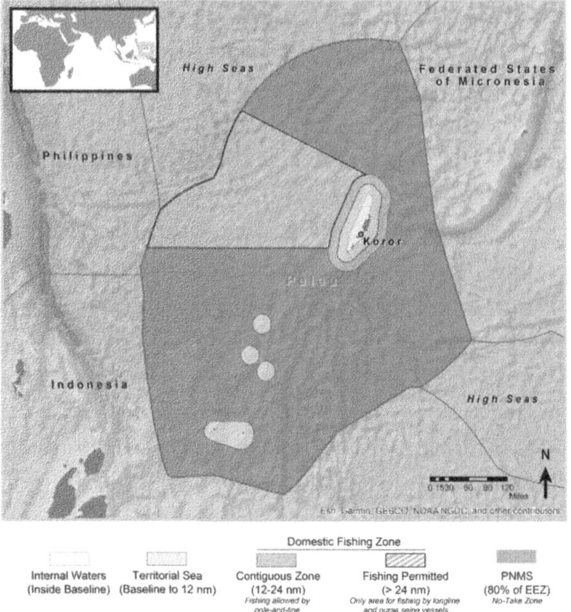

Fig. 4. Map of the PNMS and Domestic Fishing Zone (DFZ), the Revised PNMS Act, Implemented in 2020 (Dacks et al., 2020).

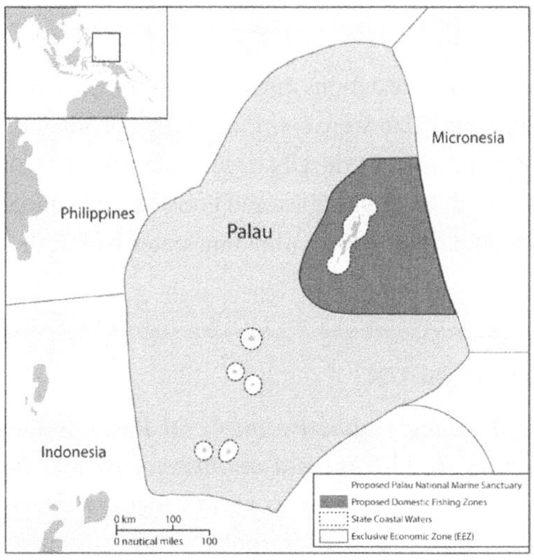

Fig. 5. Map of the PNMS and Domestic Fishing Zone (DFZ), the Original PNMS Act of 2015 (Wabnitz et al., 2018).

The cumulative effect of these efforts as noted by Palau International Coral Reef Center (PICRC) has been the creation of state and national action in the development of MPAs served as a fulcrum for biodiversity and habitat preservation (Government of Palau Bureau of Marine Resources, 2022), as noted in the 2016 PICRC technical brief (Gouezo et al., 2016):

> *Our results demonstrated that 14% of coral reef and seagrass ecosystems areas were protected; 11.2% of which were under PAN legislation. The marine habitats the most protected were channel and outer reef (>25%) and the least protected were reef flat and lagoon (<10%). Fringing and barrier reef MPAs had relatively good ecological conditions, mainly driven by the length of protection, the size of the MPAs, and the remoteness of the MPAs. Inner reef MPAs displayed good ecological condition as opposite to nearshore seagrass beds MPAs, where more than half had a score lower than 50%. For inner reef and nearshore MPAs, ecological conditions were driven by pollution caused by poor-land use.*

Primarily, this is the result of sedimentation along the edges of the mangrove forests and traditional canal systems, which continues to be under investigation by Japan International Cooperation Agency (JICA).

The Coral Triangle Initiative

This series of legislative actions and resulting improvements in biodiversity throughout Palau were coordinated by the Small Island Developing States programme at the UN to coincide with the Coral Triangle Initiative on Coral Reefs, Fisheries, and Food Security initiated in 2007 as a multilateral partnership of adjoining states within the region.

SDG 14 from the UN

'The Coral Triangle Initiative on Coral Reefs, Fisheries, and Food Security (CTI-CFF) is a multilateral partnership of six countries working together to sustain extraordinary marine and coastal resources by addressing crucial issues such as food security, climate change, and marine biodiversity. There is broad scientific consensus that the Coral Triangle represents a global epicenter of marine life abundance and diversity. Spanning only 1.6% of the planet's oceans, the Coral Triangle region is home to the highest coral diversity in the world, with 600 corals or 76% of the world's known coral species. It contains the highest reef fish diversity on the planet, with 2,500 or 37% of the world's reef fish species concentrated in the area. It is also a spawning and nursery ground for six species of threatened marine turtles, endangered fish, and cetaceans such as tuna and blue whales. These unparalleled marine and coastal living resources provide significant benefits to the approximately 363 million people who reside in the Coral Triangle, as well as billions more outside the region. As a source of food, income, and protection from severe weather events, the ongoing health of these ecosystems is critical. Recognizing the need to safeguard the region's marine and coastal resources, Indonesian President Yudhoyono inspired other leaders in the region to launch the Coral Triangle Initiative on Coral Reefs, Fisheries and Food Security (CTI-CFF) in 2007. The CTI-CFF is a multilateral partnership between the governments of Indonesia, Malaysia, Papua New Guinea, Philippines, Solomon Islands and Timor-Leste (the 'CT6')'. (Department of Economic and Social Affairs, 2009)

On 1 January 2020, the PNMS went into effect following the production of a series of technical reports by PICRC to enhance the terrestrial and biodiversity protection of regional resources (PICRC technical reports, Palau Conservation Society). Amendments to the Domestic Fishing Zone (DFZ) placement took place in 2019, allowing for the fishing zone to be placed closer to international waters and the commercial export of fish caught in that zone. It was noted at the time that this would clear the way for a more robust fishery while maintaining 80% of the territory as protected. It is unclear whether any subsequent research has been conducted on migratory patterns in the fishery since that time, given the COVID-19 pandemic. However, the change allowed the option for a domestic export market in the overall scheme of the protected area status (Carreon, 2019). This gives the characterisation of the zone the ability to be used in ways that maintain traditional usages, allow limited commercial fisheries, and encourage the development of the Belau Offshore Fisheries Industry (BOFI) and other domestic fishers while maintaining an emphasis on biodiversity and habitat.

The Spill-over Effect in Palau

Since 2020, the creation of the National Marine Sanctuary along with the network of protected areas has created a documentable biological spill-over of migratory fish species in the entire Micronesian region, and though difficult to precisely quantify at the moment researchers can begin to be documented and possibly explain increased fishery catches from regional markets in species including yellowfin tuna (*Thunnus albacares*), big eye tuna (*Thunnus obesus*) from waters in Palau that serve as a feeder for the coral triangle region. Despite overfishing of these species throughout the western central Pacific region, the PNMS area appears to be serving as a feeder population for migratory fish populations, having such an effect that a number of migratory species are not documented to be overfished (WCPFC, 2023). However, this spill-over effect is not limited to pelagic species, and an examination of reef species should and must be part of a comprehensive biodiversity and biomass survey of Palau's coastal areas. At present, no baseline surveys are available to inform decision makers (Filous et al., 2020; Vincent et al., 2020).

Muller-Karanassos et al. (2021), in the first study, suggested indigenous anthropogenic pressures to declines in reef populations that, 'fishing pressure from Koror had a significant effect on fish biomass in inner reef habitats, with a weak negative relationship observed'. This led to one observation that using MPAs as a tool for coastal repopulation is only as effective as the consistency of enforcement along coastlines (Muller-Karanassos et al., 2021).

The policy was created in tandem with regional partners. The most notable effect of cumulative years of legislation, intentionally set to sustain the protection and regeneration of biological diversity within a specific territorial area because of anthropogenic pressures, is spill-over into adjacent territories. The process of documenting biological spill-over from the PNMS is an area where international partners can and should cooperate on the overall management of fisheries resources. This will offer a mechanism to begin regional biomass quantification in larger marine sanctuaries and protected area sites. It is a change in the way we think not only of MPAs but also in the way we understand conservation on a global level by setting a precedent in the use of environmental; throughout this series of legislative decisions and conferences Palau not only met the Micronesian Challenge but exceeded the 30% protection rate it asked of participants.

Primary conservation also takes place within a framework that could be used in marine spatial planning designations as **primary marine protected areas** where legislation acts as a catalyst for larger ecological goals like biodiversity replenishment or for the development of **secondary marine protected areas** where another use has been allocated, such as a wind turbine generation area or aquaculture, that by virtue of their utility prevent commercial fishing but have the secondary and temporal effect of an ecological oasis (Grorud-Colvert et al., 2021). Multiple uses are allowed and even encouraged to support local livelihoods, power generation, recreation, and scientific research are part of the intent of having adaptive strategies coexist in a way that integrates sustainable marine ecology and sustainable human ecology in a single conceptual framework. This will be a practical buttress for SDG14 and pave the way for other pioneering legislative efforts in the areas of Integrated Multi-Trophic Aquaculture (IMTA) or regenerative aquaculture allowing them to play a larger role in association with MPAs.

The Vision of Building an Integrated Network of MPAs Following Palau

The authors recognise the need for an integrated policy idea by describing the situation in Palau and comparing this with the Greater North Sea. This idea is based on the needs of biodiversity and habitat protection, green energy, and maintaining the fishing industry.

Following Palau, the European Union can build its integrated network of MPAs where fisheries and biodiversity can both flourish through good management and integrated policy applications. Although this idea is a vision of the authors it is theoretically possible to achieve such knowledge transfer. Definitive experiments and further research in the Greater North Sea are needed to conclude. Europe and Palau share environmental protection goals through using the mechanisms of MPAs and understanding the Palauan model is useful.

REFERENCES

Ashley, M.C., Mangi, S.C. and Rodwell, L.D. 2014. The potential of offshore windfarms to act as marine protected areas – A systematic review of current evidence, *Marine Policy*, 45, 301–309. doi: 10.1016/j.marpol.2013.09.002

Carreon, B. 2019. Palau Changes Famed Marine Sanctuary Law, Allowing Commercial Exports. Available at: https://www.seafoodsource.com/news/supply-trade/palau-changes-famed-marine-sanctuary-law-allowing-commercial-exports [Accessed 10 May 2023].

Chirosca, A.M., Rusu, L. and Bleoju, A. 2022. Study on wind farms in the North Sea area, *Energy Reports*, 8, 162–168. doi: 10.1016/J.EGYR.2022.10.244

Cimino, M.A., Anderson, M., Schramek, T., Merrifield, S. and Terrill, E.J. 2019. Towards a fishing pressure prediction system for a WesternPacific EEZ, *Scientific Reports*, 9(1), 1–10. doi: 10.1038/s41598-018-36915-x

Colléter, M., Gascuel, D., Albouy, C., Francour, P., Tito de Morais, L., Valls, A. and Le Loc'h, F. 2014. Fishing inside or outside? A case studies analysis of potential spillover effect from marine protected areas, using food web models, *Journal of Marine Systems*, 139, 383–395. doi: 10.1016/j.jmarsys.2014.07.023

Dacks, R., Lewis, S.A., James, P.A.S., Marino, L.L. and Oleson, K.L.L. 2020. Documenting baseline value chains of Palau's nearshore and offshore fisheries prior to implementing a large-scale marine protected area, *Marine Policy*, 117, 103754. doi: 10.1016/J.MARPOL.2019.103754

Dame, R.F. and Kenneth, M.J. 2011. Ecology of Marine Bivalves, *Ecology of Marine Bivalves*, 284. doi: 10.1201/B11220

Department of Economic and Social Affairs. 2009. *The Coral Triangle Initiative on Coral Reefs, Fisheries and Food Security (CTI-CFF)*. Available at: https://sdgs.un.org/partnerships/coral-triangle-initiative-coral-reefs-fisheries-and-food-security-cti-cff [Accessed 1 May 2023].

Duarte, C.M., Agusti, S., Barbier, E., Britten, G.L., Castilla, J.C., Gattuso, J.P., Fulweiler, R.W., Hughes, T.P., Knowlton, N., Lovelock, C.E., Lotze, H.K., Predragovic, M., Poloczanska, E., Roberts, C. and Worm, B. 2020. Rebuilding marine life, *Nature*, 580(7801), 39–51. doi: 10.1038/s41586-020-2146-7

Evers, J. 2022. *Role of Keystone Species in an Ecosystem | National Geographic Society, National Geographic Society*. Available at: https://education.nationalgeographic.org/resource/role-keystone-species-ecosystem [Accessed 6 January 2023].

Filous, A., Friedlander, A.M., Griffin, L., Lennox, R.J., Danylchuk, A.J., Mereb, G. and Golbuu, Y. 2020. Movements of juvenile yellowfin tuna (*Thunnus albacares*) within the coastal FAD network adjacent to the Palau National Marine Sanctuary: Implications for local fisheries development, *Fisheries Research*, 230, 105688. doi: 10.1016/J.FISHRES.2020.105688.

Flanders Marine Institute (VLIZ). 2020. Underwater Noise | *European Marine Board*. Available at: https://www.marineboard.eu/underwater-noise [Accessed 10 January 2023].

Fujita, Y., Miyakuni, K. and Marino, L.L. 2022. Economic Value of Coral Reefs in Palau, 329–340. doi: 10.1007/978-981-16-6695-7_16

Gouezo, M., Koshiba, S., Otto, E.I., Olsudong, D., Mereb, G. and Jonathan, R. 2016. Ecological conditions of coral-reef and seagrass marine protected areas in Palau, *PICRC Technical Report*, 16(06). Available at: https://repository.library.noaa.gov/view/noaa/14775 [Accessed 2 May 2023].

Government of Palau Bureau of Marine Resources. 2022. *Discussions with Stephen Victor and Adelle Isechal.*

Grorud-Colvert, K., Sullivan-Stack, J., Roberts, C., Constant, V., Horta E Costa, B., Pike, E.P., Kingston, N., Laffoley, D., Sala, E., Claudet, J., Friedlander, A.M., Gill, D.A., Lester, S.E., Day, J.C., Gonçalves, E.J., Ahmadia, G.N., Rand, M., Villagomez, A., Ban, N.C., Gurney, G.G., Spalding, A.K., Bennett, N.J., Briggs, J., Morgan, L.E., Moffitt, R., Deguignet, M., Pikitch, E.K., Darling, E.S., Jessen, S., Hameed, S.O., Di Carlo, G., Guidetti, P., Harris, J.M., Torre, J., Kizilkaya, Z., Agardy, T., Cury, P., Shah, N.J., Sack, K., Cao, L., Fernandez, M. and Lubchenco, J. 2021. The MPA guide: a framework to achieve global goals for the ocean, *Science*, 373(6560). doi: 10.1126/SCIENCE.ABF0861

Halouani, G., Villanueva, C.M., Raoux, A., Dauvin, J.C., Ben Rais Lasram, F., Foucher, E., Le Loc'h, F., Safi, G., Araignous, E., Robin, J.P. and Niquil, N. 2020. A spatial food web model to investigate potential spillover effects of a fishery closure in an offshore wind farm, *Journal of Marine Systems*, 212, 103434. doi: 10.1016/J.JMARSYS.2020.103434

Halpern, B.S. 2003. The impact of marine reserves: Do reserves work and does reserve size matter? *Ecological Applications*, 13(1). doi: 10.1890/1051-0761(2003)013[0117:tiomrd]2.0.co;2

Hammar, L., Perry, D. and Gullström, M. 2016. Offshore wind power for marine conservation, *Open Journal of Marine Science*, 6, 66–78. doi: 10.4236/ojms.2016.61007

Jaiteh, V., Peatman, T., Lindfield, S., Gilman, E. and Nicol, S. 2021. Bycatch estimates from a Pacific Tuna Longline Fishery provide a baseline for understanding the long-term benefits of a Large, Blue Water Marine Sanctuary, *Frontiers in Marine Science*, 8. doi: 10.3389/FMARS.2021.720603

Jensen, H.M. and Panagiotidis, P. 2018. *Marine Regions and Subregions Under the Marine Strategy Framework Directive*. Available at: https://www.eea.europa.eu/data-and-maps/data/msfd-regions-and-subregions-1

Lindeboom, H. 2015. *Hoe konden er tonijnen van vier meter in de Noordzee leven?* Netherlads, Universiteit van Nederland. Available at: https://www.universiteitvannederland.nl/college/hoe-konden-er-tonijnen-van-vier-meter-in-de-noordzee-leven

Di Lorenzo, M., Claudet, J. and Guidetti, P. 2016. Spillover from marine protected areas to adjacent fisheries has an ecological and a fishery component, *Journal for Nature Conservation*, 32, 62–66. doi: 10.1016/j.jnc.2016.04.004

Medel, I.L. 2020. The Palau legacy pledge: A case study of advertising, tourism, and the protection of the environment, *Westminster Papers in Communication and Culture*, 15(2), 178–190. doi: 10.16997/WPCC.380

Muller-Karanassos, C., Filous, A., Friedlander, A.M., Cuetos-Bueno, J., Gouezo, M., Lindfield, S.J., Nestor, V., Marino, L.L., Mereb, G., Olsudong, D. and Golbuu, Y. 2021. Effects of habitat, fishing, and fisheries management on reef fish populations in Palau, *Fisheries Research*, 241. doi: 10.1016/J.FISHRES.2021.105996

OSPAR. 2018. 2018 Status Report on the OSPAR Network of Marine Protected Areas. Available at: https://oap-cloudfront.ospar.org/media/filer_public/50/bb/50bba6bf-4d16-4066-ad51-169d1784979d/p00730_ospar_mpa_status-report_2018.pdf [Accessed 8 February 2024].

Rice, J., Arvanitidis, C., Borja, Á., Frid, C., Hiddink, J., Krause, J., Lorance, P., Ragnarsson, Á., Sköld, M. and Trabucco, B. 2010. *Marine Strategy Framework Task Group 6 Report: Seafloor integrity*, EUR 24334 EN – Joint Research Centre, Office for Official Publications of the European Communities, Luxembourg. doi: 10.2788/85484

Rogers, A. 2019. Deep-sea fishing: would you clear-cut a forest to catch the deer? In *The Deep: The Hidden Wonders of Our Oceans and How We Can Protect Them*, Ed. A. Rogers, 1st ed., pp. 115–151. London, WILDFIRE.

Rogers, S., Casini, M., Cury, P., Heath, M., Irigoien, X., Kuosa, H., Scheidat, M., Skov, H., Stergiou, K., Trenkel, V., Wikner, J. and Yunev, O. 2010. *Marine Strategy Framework Directive Task Group 4 Report Food Webs*. doi: 10.2788/87659

Schmaltz, J. and Voiland, A. 2012. *Palau's Reefs*. Available at: https://earthobservatory.nasa.gov/images/80001/palaus-reefs [Accessed 10 May 2023].

Soukissian, T., Bondareff, J., Cummins, V., Dhanju, A., García-Soto, C., Golmen, L., Kamara, O.K., Murphy, J., Njoroge, E.M., Strati, A. and Vougioukalakis, G. 2021. Developments in renewable energy sources. In *World Ocean Assessment II*, Vol. II, pp. 321–341, New York, United Nations Division for Ocean Affairs and the Law of the Sea. Available at: https://digital.csic.es/handle/10261/321613 [Accessed 16 January 2024].

Sweeting, C.J. and Polunin, N.V.C. 2005. *Marine Protected Areas for Management of Temperate North Atlantic Fisheries: Lessons Learned in MPA Use for Sustainable Fisheries Exploitation and Stock Recovery*. Available at: https://www.researchgate.net/publication/242399251 [Accessed 31 January 2023].

The European Parliament; Council of the European Union. 2008. Directive 2008/56/EC of the European Parliament and of the Council of 17 June 2008 establishing a framework for community action in the field of marine environmental policy (Marine Strategy Framework Directive), *Official Journal of the European Union*, 51, 19–40. Available at: https://eur-lex.europa.eu/legal-content/EN/TXT/?uri=CELEX:32008L0056 [Accessed 3 January 2023].

The European Parliament; Council of the European Union. 2009. Directive 2009/28/EC of the European Parliament and of the Council of 23 April 2009 on the promotion of the use of energy from renewable sources and amending and subsequently repealing Directives 2001/77/EC and 2003/30/EC, *Official Journal of the European Union*, 1, 32–38.

The European Parliament; Council of the European Union. 2023a. *Biodiversity - GES - Marine - Environment - European Commission*. Available at: https://ec.europa.eu/environment/marine/good-environmental-status/descriptor-1/index_en.htm [Accessed 3 January 2023].

The European Parliament; Council of the European Union. 2023b. *Energy and Noise - Marine - Environment - European Commission*. Available at: https://ec.europa.eu/environment/marine/good-environmental-status/descriptor-11/index_en.htm [Accessed 10 January 2023].

The European Parliament; Council of the European Union. 2023c. *Fish and Shellfish - GES - Marine - Environment - European Commission*. Available at: https://ec.europa.eu/environment/marine/good-environmental-status/descriptor-3/index_en.htm [Accessed 3 January 2023].

The European Parliament; Council of the European Union. 2023d. *Sea-Floor Integrity - GES - Marine - Environment - European Commission*. Available at: https://ec.europa.eu/environment/marine/good-environmental-status/descriptor-6/index_en.htm

Thomsen et al. 2021. *Addressing Underwater Noise in Europe Current State of Knowledge and Future Priorities, EU Marine Board*. Available at: www.marineboard.eu

Ueki, M.F. 2000. Eco-consciousness and development in Palau, *Contemporary Pacific*, 12(2), 481–487. doi: 10.1353/CP.2000.0067

Vincent, M., Ducharme-Barth, N., Hamer, P.A., John, H., Peter, W. and Graham, P. 2020. *Stock Assessment of Yellowfin Tuna in the Western and Central Pacific Ocean*. doi: 10.13140/RG.2.2.13019.18724

Wabnitz, C.C.C., Cisneros-Montemayor, A.M., Hanich, Q. and Ota, Y. 2018. Ecotourism, climate change and reef fish consumption in Palau: benefits, trade-offs and adaptation strategies, *Marine Policy*, 88, 323–332. doi: 10.1016/J.MARPOL.2017.07.022

WCPFC. 2023. *Overview of Stock Status of Interest to the WCPFC.*

Weigel, J.Y., Mannle, K.O., Bennett, N.J., Carter, E., Westlund, L., Burgener, V., Hoffman, Z., Simão Da Silva, A., Kane, E.A., Sanders, J., Piante, C., Wagiman, S. and Hellman, A. 2014. Marine protected areas and fisheries: Bridging the divide, *Aquatic Conservation: Marine and Freshwater Ecosystems*, 24(S2), 199–215. doi: 10.1002/AQC.2514

Wijsman, J., Troost, K., Fang, J. and Roncarati, A.. 2018. *Global Production of Marine Bivalves*.

8

AQUACULTURE TEACHING AND RESEARCH IN HIGHER EDUCATION TO ADVANCE A SUSTAINABLE INDUSTRY

James Logan Sibley[a] and Matt Elliott Bell[b]

[a]*Sibley Media, USA*
[b]*University of Plymouth, UK*

ABSTRACT

In a world with over 8 billion people, ensuring sustainable food sources is paramount. This chapter explores the pivotal role of aquaculture in addressing the challenges of marine conservation and sustainable resource use. Aligned with the United Nations' Sustainable Development Goal 14, aquaculture emerges as a solution to relieve pressure on wild fish stocks and enhance food security. The chapter emphasises the rapid growth of this sector and underscores the importance of international cooperation and policies like the Global Ocean Treaty in ensuring marine biodiversity. While acknowledging the potential of aquaculture, the chapter delves into environmental concerns surrounding fishmeal and fish oil in feed. It advocates for innovative technologies and ingredients to establish a circular bioeconomy. The significance of higher education in advancing sustainable aquafeed technology, breeding, and genetics is highlighted, with a discussion on milestones achieved by experts like Dr John E. Halver and Professor Simon J. Davies. Examining technological advances, the chapter explores molecular genetics, transgenics, and gene editing, particularly CRISPR biosciences, as transformative tools for enhancing aquaculture productivity and sustainability. Environmental impacts are addressed, proposing solutions such as Recirculating Aquaculture Systems (RAS) and Multitrophic Aquaculture Systems (MTA) to minimise ecological footprints. Throughout, there is a strong emphasis on the integral role of research and education in fostering sustainable aquaculture practices. The chapter advocates for specialised courses and programs in higher education to prepare the next generation for the challenges and opportunities in aquaculture, ensuring its contribution to global food security and environmental stewardship.

Keywords: Biodiversity; CRISPR; MTA; high education; fish nutrition; sustainable aquaculture

INTRODUCTION

The world population continues to grow, and the demand for seafood is increasing, placing pressure on our oceans and marine ecosystems. The United Nations' Sustainable Development Goal 14 (SDG14) aims to 'conserve and sustainably use the oceans, seas and marine resources for sustainable development'. One way to meet this mandate is through aquaculture, fish, and shrimp farming, which have the potential to provide a sustainable source of seafood while reducing pressure on wild fish stocks. Aquaculture has been growing rapidly in recent years and is expected to continue to do so in the coming decades. According to the Food and Agriculture Organization of the United Nations (FAO, 2020), aquaculture production increased from 18 million tonnes in 2000 to 82 million tonnes in 2020, making it the fastest-growing food production sector in the world. The role of aquaculture in the 2020s has an ever-increasing importance. The global population as of November 2022 reached 8 billion (UN, 2022). With fisheries already stretched to capacity and a need for sustenance, life below water is becoming an ever-diminishing source of blue foods (food from aquatic resources) under the current lack of legislation and policy within international waters. Therefore, the recently signed agreement of the Global Ocean Treaty in 2023 aimed to protect marine biodiversity within international waters is a significant advance towards more productive oceans by 2030. However, worldwide cooperation and the need for co-responsibility are required to maintain this target. Aquaculture must become a sustainable source of blue foods to mitigate humanity's impact on the ocean. To highlight this urgent necessity, The White House issued a statement in March 2022 regarding a National Security Memorandum on Strengthening the Security and Resilience of United States Food and Agriculture that included a robust aquaculture strategy. In 2023, the British Veterinary Association launched a new directive statement on UK sustainable finfish aquaculture, advocating for more research into the welfare requirements of fish and the impact of aquaculture on the environment (BVA April 2023).

While aquaculture has the potential to provide a sustainable source of seafood, there are also environmental and social challenges that need to be addressed. One of the main environmental challenges associated with aquaculture is the use of fishmeal and fish oil as feed for farmed fish. These ingredients are typically sourced from wild fish stocks, which can lead to overfishing and depletion of marine ecosystems. To address this challenge, scientists and the aquaculture industry have been developing and using novel technologies and ingredients to replace fishmeal and fish oil within aquafeeds and to advocate a circular bioeconomy for a sustainable aquaculture industry (Colombo et al., 2022). On an EU level, projects such as Neptunus focus on sustainable development towards waste management through the implementation of a circular economy strategy, eco-labelling and defining eco-innovation protocols from production to consumption (Neptunus Project, 2018). The EU SAFE project aims to create sustainable aquaculture facilities. Both projects are vital in creating bio-circularity and reducing humanity's impact towards life below water. As such, aquaculture fish and shrimp farming have the potential to meet the SDG14 mandate of conserving and using our oceans and marine resources for sustainable development. By using novel technologies and ingredients to mitigate the use of fishmeal and fish oil in fish feed, scientists and the aquaculture industry can help ensure the security of this sector by developing superior aquafeeds for fish and shrimp based on precise nutritional requirements to meet higher efficacy. Other key areas for development include the choice of species with traditional fish like Atlantic salmon (*Salmo salar*) and tilapia being improved by genetics and breeding technology and the emergence of various candidate species (e.g. grouper, pompano, mahi mahi, cobia, snakehead, and tuna) in parts of the world.

The effectiveness of selective breeding programmes has produced fish with superior growth characteristics and disease resistance. There has been a plethora of research in leading universities concerning fish health and the use of modern therapeutics and vaccines. However, continued research and innovation are needed to address the environmental and social challenges associated with aquaculture and to ensure its long-term sustainability.

NUTRITION AND AQUAFEED TECHNOLOGY

Higher education has a significant role to play in advancing the field of fish and shrimp nutrition, which is critical for the sustainability of the aquaculture industry.

The future of aquatic protein: Implications for protein sources in aquaculture diets was recently stated by Hua et al. (2019). The potential for commodities like plant by-product meals from crops, e.g. soybean, rapeseed meal, corn gluten meal, beans, and peas are being realised. Other sources of high-quality proteins originate from the animal rendering sector with such ingredients as poultry by-product, feather meal and blood meals being utilised. Through the platform of *International Aquafeed* Magazine, Professor Davies has as editor contributed to the global dissemination of such developments over 25 years. Recent trends in the generation of novel aquafeed ingredients have been based on academic inspired research and collaboration with industry. One such technology is microbial fermentation, which involves using bacteria or yeast to convert plant-based ingredients into a protein-rich feed that can be used in aquaculture. This technology has been used successfully to produce single-cell protein (SCP), which can replace up to 100% of the fishmeal in fish feed without compromising fish growth or health (Davies and Wareham, 1988; Yang et al., 2023). Another approach is the use of insect meal, which involves using insects such as black soldier fly larvae as a source of protein and fat for fish feed. Insect meal has been shown to be a viable alternative to fishmeal, with some studies reporting comparable growth and survival rates in fish-fed insect meal-based diets (Bolton et al., 2021). Over the last 50 years, we have witnessed many significant milestones in the aquaculture sector. Fig. 1 depicts the timeline for the specific developments in fish nutrition and feed technology applicable to the specialised discipline of the book editor Professor Simon J. Davies. The main fundamentals of fish nutrition were established in the 1960s by the pioneering work of Dr John E. Halver based at the University of Washington, Seattle, USA in which the major nutritional requirements for many species and in particular salmonid fish were determined (Hardy et al., 2021). Halver was considered the 'father of fish nutrition' and initiated the discipline at the university level leaving a global legacy. Dr Halver was a mentor for S.J. Davies, illustrating this continuum of academia in the aquaculture sector. Fig. 1 depicts a schematic overview of the major milestones and achievements over six decades. This is where Professor Simon J. Davies, a leading expert in fish nutrition and aquaculture, has paved the way for teaching the future of tomorrow's students and guiding those currently within positions of responsibility for over 40 years. In this capacity, Professor Davies has mentored many PhD and master's students as well as a myriad of undergraduate students in aquaculture nutrition and related fields to his esteem. He has forged the bridging of the generations and embedded the concept of sustainable fish feed since 1987 into his teaching and research. One of the most notable areas has been his pioneering efforts to seek alternative protein and energy rich ingredients for fish diets for a wide range of species such as rainbow trout (*Oncorhynchus mykiss*), Atlantic salmon (*Salmo salar*), tilapia (*Oreochromis* spp.), catfish (*Clarius* spp.), and carp (*Cyprinidae* spp.). The authors recommend publications by: Davies et al. (2021), Davies and

Gouveia (2010), Adeoye et al. (2021) which undertook efficacy feeding trials testing different animal and plant by-products as well as novel feed ingredients for aquaculture.

One of the key roles of higher education in aquaculture nutrition is to conduct research to improve our understanding of the nutritional requirements of farmed species of fish and shrimp. This includes studying the composition and digestibility of different feed ingredients, developing new feed formulations that meet the nutritional needs of fish and shrimp, and identifying and mitigating nutritional deficiencies and imbalances. Professor S.J. Davies forged his track record of fish nutrition work at the University of Plymouth (1986–2015) inspiring and mentoring many generations of students at undergraduate, masters, PhDs, and postdoctoral fellows. He has witnessed the milestones outlined in Fig. 1 and been actively involved in the scientific work, especially in the applied aspect of evaluating novel feed ingredients. One area has been his work on processed animal by-products (PAPS) for use in aquafeeds during the BSE crisis of the late 1990s where these were banned in European feeds including for fish. His perseverance to undertake validation and efficacy trials acted as supporting evidence for their reintroduction by the EU with changes in legislation allowing poultry

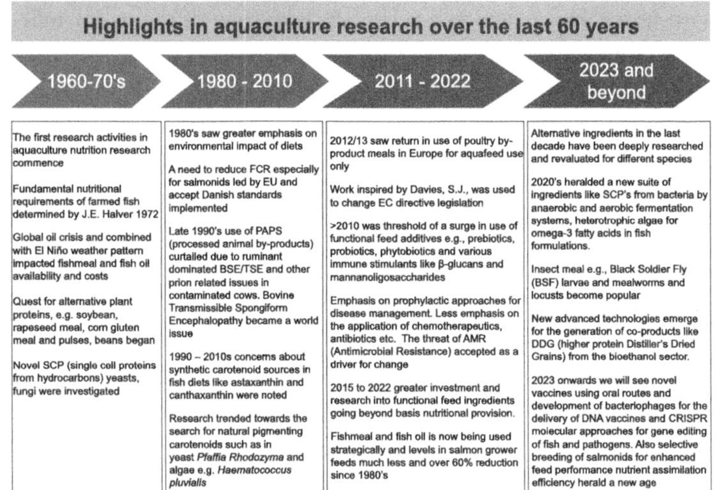

Fig. 1. Milestones in the Evolution of Aquaculture Nutrition, Research, and Development Over the Last Six Decades. Created Using Draw.io.

by-product use. Specific work entailed conducting digestibility trials on marine fish such as seabream, seabass, and turbot (Davies et al., 2009). Working with leading practitioners and experts in this field Professor Davies led studies on fish and with his associates, his team produced a comprehensive review on PAPS in aquaculture with the culmination of the comprehensive review by Woodgate et al. (2022). Other notable work has addressed the importance of research to raise fish quality. A case in point is to develop diets to enhance the omega-3 lipid profile of tilapia that had been criticised for having high levels of omega-3 content due to dietary plant oils. A study by Davies et al. (2022) conducted pioneering studies to use a novel marine microalgal source to modulate the fillet muscle status of tilapia to create a healthier product for human consumption.

The use of alternative protein and lipid sources in fish feed, as well as the development of precision feeding technologies that allow farmers to optimise their feeding practices and reduce waste. The Norwegian University of Science and Technology has been conducting research on the use of insect meal as a protein source in fish feed. In a recent study, researchers at the university found that insect meal could replace up to 50% of the fishmeal in fish feed without compromising fish growth or health (Lock et al., 2016). Research conducted at the university, researchers at the university developed a functional feed containing prebiotics and probiotics that improved the gut health and disease resistance of Atlantic salmon (Castro et al., 2020). Higher education institutions around the world are engaging with SDG14 by incorporating teaching on fish disease and health with respect of the microbiome and functional feed additives into their programs. Another example is the work of Hoseinifar et al. (2017) at the University of Agricultural Sciences and Natural Resources, Gorgan, Iran that have reviewed the role of pre-pro- and synbiotics in maintaining gut health in fish.

DEVELOPMENTS IN BREEDING AND GENETICS

Breeding and genetics have been extensively studied and applied in aquaculture to improve fish growth, health, and adaptability to various environments. This has led to the development of new strains of fish with improved performance and higher yields, which can contribute to sustainable aquaculture and food security such as the work on developing YY male tilapia (Li et al., 2022). Selective breeding has been used for many years to improve the growth and disease resistance of Atlantic salmon, one of the most economically important species in the aquaculture industry. Studies have shown that by selecting and breeding fish with higher growth rates and lower susceptibility to diseases, it is possible to increase the productivity of salmon farms and reduce the use of antibiotics and other chemicals. For example, a study conducted in Norway demonstrated that over a period of 25 years, selective breeding of Atlantic salmon resulted in a 50% increase in growth rate and a 75% reduction in mortality rates due to infectious salmon anaemia (ISA) (Gjedrem, 2012).

Tilapia (*Oreochromis* spp.) is another popular species in the aquaculture industry, and genomic tools have been used to improve its breeding and performance (Li et al., 2022). In a study published in *Nature Communications* in 2018, researchers used genomic markers to identify genes associated with growth and disease resistance in tilapia and used this information to develop a genomic selection model for breeding. This approach led to a 15% increase in growth rate and a 40% reduction in mortality due to bacterial infections (Palaiokostas et al., 2018). Hybridisation, or crossbreeding of different fish species, has also been used in aquaculture to produce fish with desirable traits such as fast growth and disease resistance. In the case of catfish (*Clarias* spp.), a study conducted in Vietnam found that crossing two different strains of catfish resulted in hybrids with faster growth and higher survival rates than either parent strain (Chaivichoo et al., 2020).

ADVANCES IN MOLECULAR GENETICS, TRANSGENICS AND GENE EDITING (CRISPR) BIOSCIENCES

The application of genomics and transgenics in aquaculture has also gained significant attention due to its potential to improve productivity and promote sustainable development in the industry. Next-generation sequencing (NGS) technologies have revolutionised genomic research by providing high-throughput, cost-effective, and rapid sequencing of DNA and RNA molecules. In aquaculture, NGS has been used to investigate the genetic basis of important traits such as growth, disease resistance, and stress tolerance in various fish species. For instance, researchers at the University of Stirling used NGS to sequence the genome of Atlantic salmon and identify genes associated with traits related to growth and disease resistance. Houston and Macqueen (2018) have comprehensively reviewed the potential of genetics in the 21st century.

Transgenics involves the introduction of foreign DNA into an organism's genome to alter its genetic makeup. In aquaculture, transgenic fish have been developed to enhance growth rates, disease resistance, and tolerance to environmental stressors. The author J. Sibley has highlighted the molecular and cell biology mechanisms and their application in the production of transgenic fish and the specific interests in the use of CRISPR and used in enabling the potential of gene editing for expressing desirable traits and modern disease prevention and treatment by the development of bacteriophagic type agents for aquaculture. These can be viewed positively to provide more efficient fish leading to enhanced sustainability as well as economic gain. This has been the basis of the success story of the US company AquaBounty Technologies in their pioneering work to produce the iconic AquAdvantage Atlantic salmon product. AquAdvantage salmon exhibit nearly doubled growth rates due to transgenic growth modifier inclusions from coho salmon and poutfish. These chimeric salmon were engineered decades before CRISPR-Cas complexes were discovered. AquAdvantage salmon were the first genetically engineered animals approved for human consumption in the United States and Canada. This approval has been subject to much controversy (FDA, 2016) and AquaBounty is now an established enterprise based in the Midwest of the United States in the state of Indiana. A practical example of CRISPR in Atlantic salmon production would be the ability to produce Germ Cell-Free (GCF) fish as an effective method of sterilisation without the need to produce triploid fish with their specific health and growth-related downsides (Madaro et al., 2022). Moreover, GCF salmon are nutritionally indistinguishable from wild Atlantic salmon and remain immature, improving growth rates and reducing the likelihood of escapee male competition with wild populations. The scientific challenge with GCF salmon is that, without germ cells, they cannot be produced at the scale and volume necessary to support vigorous sea farm operations. GCF salmon require embryonic CRISPR-Cas9 dnd knockout, resulting in development without gonads (Güralp et al., 2020). In order to viably scale GCF salmon aquaculture, broodstock must be fertile while their offspring must be GCF. The authors hypothesised that crispant embryos could be rescued and develop gonads while retaining the dnd-KO in their germline. Rescued GCF salmon brood-stock could provide the scale necessary for GCF ocean aquaculture without compromise.

In a review article, Dunham and Su (2020) described genetically engineered fish and their potential impacts on aquaculture, biodiversity, and the environment. The use of genetically modified fish in aquaculture has raised concerns about their potential impact on wild fish populations and the environment. In this regard, the use of genetically modified fish should be balanced with sustainable aquaculture practices that prioritise environmental stewardship has been eloquently advocated by J. Sibley (2022/2023) in social media platforms. The author has used his expert skills, training, and academic studies of cell and genetics specialisations in biology at Northeastern University, Boston, USA to achieve a translational scientific approach to employ in the aquaculture biosciences domain.

AQUACULTURE AND THE ENVIRONMENT

Moreover, higher education institutions have also been involved in developing sustainable aquaculture practices to minimise the impact on the environment. The University of California, Santa Cruz, has been involved in research to develop sustainable aquaculture practices to produce seaweed, which can be used as a food source for humans and animals. The research team has developed a system that uses wastewater from fish farms to grow seaweed, which helps to reduce the environmental impact of fish farming.

By continuing to support research and education in sustainable aquaculture practices, higher institutions can help to ensure that fisheries and aquaculture are sustainable and productive for generations to come. In terms of teaching sustainable aquaculture, universities are increasingly incorporating novel technologies and engineering solutions, such as Recirculating Aquaculture Systems (RAS) and Multitrophic Aquaculture Systems (MTA), into their programs. RAS is a technology that recirculates water to reduce environmental impact, while MTA integrates multiple species to create a more balanced ecosystem. For instance, Cottrell et al. (2021) emphasised that we need to rethink trophic levels in aquaculture policy and consider the ecological integration of complex food webs into the production system.

In response to the growing concerns over the environmental impact of salmon farming, many farms are taking innovative measures to mitigate these effects, with the inclusion of kelp farming being one of the most intriguing strategies. This symbiotic practice, known as integrated multi-trophic aquaculture (IMTA), cultivates salmon alongside species like kelp that can absorb and metabolise the waste produced by the salmon. Kelp, a fast-growing seaweed, consumes the nitrogenous waste and carbon dioxide produced by the salmon, effectively turning these potentially harmful by-products into a valuable resource for growth. As a result, the need for chemical waste treatment is substantially reduced, and the kelp itself can be harvested and used for various purposes, such as food, biofuel, and even cosmetic products.

ROLE OF ACADEMIA IN PROMOTING NOVEL SOLUTIONS, SOCIETAL AND POLICY

Partnerships are of paramount importance between NGOs and Academica. For example, the Global Aquaculture Alliance, an international non-governmental organisation, partnered with the University of Stirling in Scotland to develop a certification programme for sustainable aquaculture practices. This programme aims to ensure that aquaculture products are produced in an environmentally and socially responsible manner. The Food and Agriculture Organization of the United Nations (FAO) has established the Blue Growth Initiative, which aims to promote sustainable economic growth in the aquatic sector. Universities and academic institutions are playing an important role in advancing SDG14 by incorporating novel technologies and engineering solutions, such as RAS and MTA, into their programs. Collaborations between universities and organisations such as the Global Aquaculture Alliance and the FAO further support sustainable aquaculture practices on a global scale. SDG14 focuses on 'life below water', which includes ensuring sustainable fisheries, protecting marine ecosystems, and reducing pollution in the oceans. For instance, researchers at the Norwegian University of Science and Technology have developed a sustainable aquaculture system that uses waste from fish farming to cultivate microalgae as a feed source for fish (Mandal et al., 2020). This approach can reduce the environmental impact of aquaculture by reducing the use of traditional fishmeal and reducing waste.

While studying at the University of Galway and the Carna marine station, in Ireland as part of the author's (M. E. Bell) placement year supervised by Dr Alex Wan, the author worked on several projects. The author aimed to comprehend the wider aquaculture remit during his short time in Ireland at the University of Galway as a placement student. During this time, M.E. Bell supported the research initiatives on farmed fish and shellfish, including practices to conserve threatened species and stock enhancement schemes, such as the European Lobster (*Homarus gammarus*). While on campus, at the Ryan Institute, he worked on policy papers for the European Commission under the EU Neptunus Program (EAPA_576/2018-NEPTUNUS) and helped to set up the Kickstarter meeting for a then newly funded EU project SAFE, concerning sustainable aquaculture (SAFE, 2022), alongside laboratory work such as proximate analysis and blood assays. Prior to this experience, the author arrived at the pre-determined notion that aquaculture (mainly salmon farming) was detrimental to the environment, unethical and unsustainable, which is portrayed by the news and is the public consensus (Osmundsen and Olsen, 2017). However, at the end of the author's time at the University of Galway, the author has found a new respect and an open awareness of the benefits towards aquaculture for attaining global food security. This has been a driver in the authors' commitment towards promoting and disseminating the benefits of this industry through opportunities and guidance of Professor Davies.

Additionally, through the actions of Sibley, Davies, and Bell, a more educated and enlightened viewpoint is being projected to provide positive images for this industry whilst respecting the problems and endeavouring to solve the issues remaining. In this respect, all authors are sensitive to this issue and J. Sibley has made stringent efforts to advance public understanding of modern seafood production through aquaculture. His internship experience has embraced the seafood industry and in particular retail and the consumer where he has emphasised key aspects of governance, societal involvement, and engagement. This necessitates establishing viable pathways to impact, empowering stakeholders, retailers, and consumers involved in the seafood chain. The goal would be to ensure confidence and capacity building and to embed the ethos of SDG14 linking to the higher education sector.

CONCLUSION

Research in higher education has played a significant role in advancing aquaculture using breeding and genetics, disease, health, nutrition, and feed technology. These have led to the development of new strains of fish with improved performance contributing to the growth and development of the aquaculture industry. The huge strides have enabled much better welfare standards to be adopted in many advanced countries and the mechanism to bring this awareness and knowledge to developing countries. This will mean a requirement to offer specialised higher educational courses and programs at different levels ranging from diploma, undergraduate, and master's degrees to doctoral programs and beyond. The opportunities for junior apprenticeship schemes and vocational training will also provide an invaluable experience to guide the younger generation of students to employment and careers and to potentially enter both further and higher education relevant to sustainable and secure aquaculture practices.

REFERENCES

Adeoye, A.A., Akegbejo-Samsons, Y., Fawole, F.J., Olatunji, P.O., Muller, N., Wan, A.H.L. and Davies, S.J. 2021. From waste to feed: dietary utilisation of bacterial protein from fermentation of agricultural wastes in African catfish (*Clarias gariepinus*) production and health, *Aquaculture*. doi: 10.1016/j.aquaculture.2020.735850

Bolton, C.M., Muller, N., Hyland, J., Johnson, M.P., Valente, C.S., Davies, S.J. and Wan, A.H.L. 2021. Black soldier fly larval meal with exogenous protease in diets for rainbow trout (*Oncorhynchus mykiss*) production meeting consumer quality, *Journal of Agriculture and Food Research*. doi: 10.1016/j.jafr.2021.100232

British Veterinary Association (BVA). 2023. BVA, Welfare of farmed salmon. *BVA*. Available at: bva.co.uk/aquaculture [Accessed on 19 April 2023].

Colombo, S.M., Roy, K., Mraz, J., Wan, A.H.L., Davies, S.J., Tibbetts, S.M., Øverland, M., Francis, D.S., Rocker, M.M., Gasco, L., Spencer, E., Metian, M., Trushenski, J.T. and Turchini, G.M. 2022. Towards achieving circularity and sustainability in feeds for farmed blue foods, *Reviews in Aquaculture*, 1–27. doi: 10.1111/raq.12766

Castro, C., Lima, C.M., Paula, J.R., Machado, M., Silva, T.S. and Peixoto, M. J. 2020. Functional feed improves gut health and disease resistance of Atlantic salmon (*Salmo salar*) fed a low fish meal diet, *Aquaculture*, 529, 735643.

Chaivichoo, P., Koonawootrittriron, S., Chatchaiphan, S., Srimai, W. and Na-Nakorn, U. 2020. Genetic components of growth traits of the hybrid between ♂North African catfish (*Clarias gariepinus* Burchell, 1822) and ♀bighead catfish (*C. macrocephalus* Günther, 1864), *Aquaculture*, 521. doi: 10.1016/j.aquaculture.2020.735082

Cottrell, R.S., Metian, M., Froehlich, H.E., Blanchard, J.L., Sand Jacobsen, N., McIntyre, P.B., Nash, K.L., Williams, D.R., Bouwman, L., Gephart, J.A., Kuempel, C.D., Moran, D.D., Troell, M. and Halpern, B.S. 2021. Time to rethink trophic levels in aquaculture policy, *Reviews in Aquaculture*. doi: 10.1111/raq.12535

Davies, S.J. and Gouveia, A. 2010. Response of common carp fry fed diets containing a pea seed meal (*Pisum sativum*) subjected to different thermal processing methods, *Aquaculture*, 305(1–4), 117–123. doi: 10.1016/j.aquaculture.2010.04.021

Davies, S.J. and Wareham, H. 1988. A preliminary evaluation of an industrial single cell protein in practical diets for tilapia (*Oreochromis mossambicus* Peters). *Aquaculture*. doi: 10.1016/0044-8486(88)90053-1

Davies, S.J., El-Haroun, E.R., Hassaan, M.S. and Bowyer, P.H. 2021. A Solid-State Fermentation (SSF) supplement improved performance, digestive function, and gut ultrastructure of rainbow trout (*Oncorhynchus mykiss*) fed plant protein diets containing yellow lupin meal, *Aquaculture*, 545. doi: 10.1016/j.aquaculture.2021.737177

Davies, S.J., Gouveia, A., Laporte, J., Woodgate, S.L. and Nates, S. 2009. Nutrient digestibility profile of premium (category III grade) animal protein by-products for temperate marine fish species (European sea bass, gilthead sea bream and turbot), *Aquaculture Research*. doi: 10.1111/j.1365-2109.2009.02281.x

Davies, S.J., Roderick, E., Brudenell-Bruce, T., Bavington, C.D., Hartnett, F., Hyland, J., de Souza Valente, C. and Wan, A.H.L. 2022. Delivering a nutritionally enhanced tilapia fillet using a pre-harvest phase Omega-3 Thraustochytrids Protist Enriched Diet, *European Journal of Lipid Science and Technology*. doi: 10.1002/ejlt.202100153

Dunham, R.A. and Su, B. 2020. Genetically engineered fish: potential impacts on aquaculture, biodiversity, and the environment. In *GMOs. Topics in Biodiversity and Conservation*, Eds A. Chaurasia, D.L. Hawksworth and M. Pessoa de Miranda, Vol. 19, Cham, Springer. doi: 10.1007/978-3-030-53183-6_11

FAO. 2020. Impacts of climate change on fisheries and aquaculture. Synthesis of current knowledge, adaptation, and mitigation options. In *FAO Fisheries and Aquaculture Technical Paper 627. Food and Agriculture Organization of the United Nations*. Food and Agriculture Organisation.

FDA. 2016. *FDA Import Alert 99-40 "Genetically Engineered (GE) Salmon"*. FDA. 2016-01-29. Archived from the original on 2016-01-31.

Gjedrem, T. 2012. Genetic improvement for the sustainable development of aquaculture, *Frontiers in Genetics*. doi: 10.1016/J.AQUACULTURE.2012.03.003

Güralp, H., Skaftnesmo, K.O., Kjærner-Semb, E., Straume, A.H., Kleppe, L., Schulz, R.W., Edvardsen, R.B. and Wargelius, A. 2020. Rescue of germ cells in *dnd* crispant embryos opens the possibility to produce inherited sterility in Atlantic salmon, *Scientific Reports*, 10, 18042. doi: 10.1038/s41598-020-74876-2

Hardy, R.W., Kaushik, S.J. and Mai, K. 2021. Fish nutrition-history and perspectives. In *Fish Nutrition*, Eds R. Hardy and S.J. Kaushik, 4th ed. Elsevier Academic Press.

Hoseinifar, S., Dadar, M. and Ringø, E. 2017. Modulation of nutrient digestibility and digestive enzyme activities in aquatic animals: the functional feed additives scenario, *Aquaculture Research*, 48(8), 1–14. doi: 10.1111/are.13368

Hua, K., Cobcroft, J.M., Cole, A., Condon, K., Jerry, D.R., Mangott, A., Praeger, C., Vucko, M.J., Zeng, C., Zenger, K. and Strugnell, J.M. 2019. The future of aquatic protein: implications for protein sources in aquaculture diets, *One Earth*. doi: 10.1016/j.oneear.2019.10.018

Houston, R. D. and Macqueen, D. J. 2018. Atlantic salmon (*Salmo salar L.)* genetics in the 21st century: taking leaps forward in aquaculture and biological understanding, *Animal Genetics*, 50(1), 3–14. doi: 10.1111/age.12748

Li, M., Liu, X., Lu, B., Sun, L. and Wang, D. 2022. Sexual plasticity is affected by sex chromosome karyotype and copy number of sex determiner in tilapia, *Aquaculture*. doi: 10.1016/j.aquaculture.2022.738664

Lock, E.R., Arsiwalla, T. and Waagbø, R. 2016. Insect larvae meal as an alternative source of nutrients in the diet of Atlantic salmon (*Salmo salar*) postsmolt, *Aquaculture Nutrition*, 22, 1202–1213. doi: 10.1111/anu.12343

Madaro, A., Kjøglum, S., Hansen, T., Fjelldal, P.G. and Stien, L.H. 2022. A comparison of triploid and diploid Atlantic salmon (*Salmo salar*) performance and welfare under commercial farming conditions in Norway, *Journal of Applied Aquaculture*, 34(4), 1021–1035. doi: 10.1080/10454438.2021.1916671

Mandal, S.C., Nordgreen, A., and Mathiesen, S.D. 2020. Sustainable aquaculture: a review of carbon and nutrient footprint of integrated multi-trophic aquaculture systems, *Reviews in Aquaculture*, 12(1), 132–148.

Neptunus Project. 2018. *Neptunus*. Available at: NEPTUNUS (https://neptunus-project.eu/)

Osmundsen, T.C. and Olsen, M.S. 2017. The imperishable controversy over aquaculture, *Marine Policy*. doi: 10.1016/j.marpol.2016.11.022

Palaiokostas, C., Kocour, M., Prchal, M. and Houston, R. D. 2018. Accuracy of genomic evaluations of juvenile growth rate in common carp (*Cyprinus carpio*) using a 2000 fish commercial breeding population, *Genetics Selection Evolution*. doi: 10.3389/fgene.2018.00082

SAFE. 2022. *SAFE*. Available at: http://projectsafe.eu/about/

Sibley, J. Faces of the Future. 2022. Why James Sibley is backing CRISPR's impact on aquaculture; Gene-editing tools such as CRISPR have huge potential to improve the sustainability and profitability of the aquaculture industry, according to James Sibley – an undergrad with ambitions to be a star of the sector. The Fish Site 17 February 2023.

The White House. 2022. *National Security Memorandum on Strengthening the Security and Resilience of United States Food and Agriculture*. The White House. Available at: https://www.whitehouse.gov/briefing-room/presidential-actions/2022/11/10/national-security-memorandum-on-on-strengthening-the-security-and-resilience-of-united-states-food-and-agriculture/

UN. 2022. *World Population to Reach 8 Billion on 15 November 2022*. United Nations, Department of Social and Economic Affairs. Available at: https://www.un.org/en/desa/world-population-reach-8-billion-15-november-2022#:~:text=World%20population%20to%20reach%208%20billion%20on%2015%20November%202022

Woodgate, S.L., Wan, A.H.L., Hartnett, F., Wilkinson, R.G. and Davies, S.J. 2022. The utilisation of European processed animal proteins as safe, sustainable and circular ingredients for global aquafeeds, *Reviews in Aquaculture*. doi: 10.1111/raq.12663

Yang, P., Li, X., Song, B., He, M., Wu, C. and Leng, X. 2023. The potential of *Clostridium autoethanogenum*, a new single cell protein, in substituting fish meal in the diet of largemouth bass (*Micropterus salmoides*): growth, feed utilization and intestinal histology, *Aquaculture and Fisheries*. doi: 10.1016/j.aaf.2021.03.003

9

REALISING SUSTAINABLE DEVELOPMENT: HARMONISING PEOPLE, PLANET AND PROFIT IN LARGE-SCALE DEVELOPMENT PROJECTS

Davy van Doren

MatureDevelopment B.V., The Netherlands

ABSTRACT

Although large-scale construction projects can stimulate economic development, they can also cause unanticipated environmental stress. In addition, there are indications that such projects can collide with local cultural structures and create negative social impacts. With a focus on *Building with Nature* – an initiative towards sustainable hydraulic engineering – this chapter illustrates how nature conservation can be integrated into the daily operation of large-scale construction projects. Also, some insights are presented on the effects of voluntary green behaviour, particularly about challenges and benefits associated with enforcing corporate responsibility. The chapter concludes with a discussion on the role of integrative systematic approaches in analysing the complexity related to multi-stakeholder involvement for the embodiment of SDG14 Life Below Water. Also, some arguments are provided on the value of intergenerational knowledge exchange – linking expertise and experience of industry representatives with innovative concepts from higher education actors – for realising goals linked to sustainable development embracing future generations.

Keywords: Large-scale construction projects; hydraulic engineering; nature conservation; sustainability legislation; corporate responsibility; stakeholder integration and collaboration

INTRODUCTION

Within human nature, a general tension exists between the urge for preservation and one for development. It is the construction of things we desire versus the conservation of what we seek to keep. Regarding our natural world, there is a dominant concern that – in light of the current prevalent political landscapes and economic structures globally – active and rigid protection against negative human influence is necessary. Particularly in aquatic ecosystems, sustainable development is often considered from a conservation perspective. Over this largely invisible world below the surface of the visible, incapacity and ignorance have ruled for long.

Since the publication of the Brundtland Report, sustainable development has been widely regarded as a form of development that meets the needs of the present but without compromising those of the future (The Brundtland Commission, 1987). Conceptually, it can be argued that sustainable development has not changed much in recent decades. Pragmatically, the debate on how sustainable development should be implemented or can be achieved has evolved dramatically.

This chapter addresses some fundamental criteria influencing operational success and the degree of sustainability achieved in large-scale construction projects. Using a case study in Abu Dhabi, the reflection mainly concerns collaboration and stakeholder integration in the context of development projects. On the one hand, there are illustrations that, despite increased awareness of corporate responsibility and the green economy, economic progress has not always gone hand in hand with issues associated with sustainable development. On the other hand, there is also evidence that the tension within the multi-dimensionality of sustainable development can be resolved in practice.

Also, some arguments are provided on the potential function and value of collaboration and intergenerational knowledge exchange – partly based on Freire's notion of critical pedagogy, as well as on gathered experience in linking the creative forces from higher education platforms with those of existing industries – in achieving set goals for sustainable development.

SUSTAINABLE DEVELOPMENT IN PRACTICE – BUILDING WITH NATURE

Challenges of Aquatic Ecosystem Monitoring

When the World and Palm Islands rose from the sediments of the Persian Gulf, it demonstrated the power of imagination and will becoming materialised. These artificial structures considered significant feats of engineering art, symbolically placed the United Arab Emirates on the map of the global economy. However, with the focus of new projects often centred on the potential benefits they bring once initiated or completed, many of the negative impacts associated with development are not always well anticipated. In the wake of these creations and following voices of praise, more critical viewpoints arose regarding the impact of significant dredging and construction activity on the quality of the natural environment (Mansourmoghaddam et al., 2022).

Not far from these newly created lands before the coast of Dubai, the construction of Khalifa Port and Industrial Zone in 2008 by Boskalis – a Dutch and globally operating hydraulic engineering company – was initiated to overcome the constraints of the old city port Mina Zayed and lay the grounds for major diversification in relation to real estate, tourism and infrastructure in Abu Dhabi (Boskalis, 2012). However, it was also a greenfield project of *Building with Nature* – a programme managed by the network Ecoshape – to introduce a more inclusive approach to hydraulic engineering (de Vriend et al., 2015). In addition to members of industry, *Building with Nature* generally has a strong involvement in higher education, including core member Wageningen Marine Research (Wageningen University) and various network partners, including TU Delft and University Twente.[1] Such actors from higher education also potentially provide rich prospects for student engagement to appreciate the challenges of SDG14 'Life Below Water' within this context of marine system conservation and living resource management.

For corals, it is well known that silt pollution can have a severe negative impact on growth (Rogers, 1990). Many studies have already measured the adverse physical, physiological, behavioural, developmental, and ecological responses of corals to increased concentration, duration, and frequency of sediment level exposure (Tuttle et al., 2020). *Building with Nature* was intended to monitor in real-time such effects of dredging activity, to report on relationships between dredging activity and aquatic biospheres, and, as such, to control the overall nature and intensity of hydraulic engineering. In contrast to more traditional reactive approaches – generally focused on minimising negative impacts and compensating for potential residual effects of infrastructural projects – *Building with Nature* was aimed to proactively utilise natural processes and provide opportunities for nature as part of the infrastructure development process (de Vriend et al., 2015).

A Laboratory for Environmental Conservation

Not unexpectedly, the initiation of environmental monitoring was challenging. From a personal perspective, the transfer of the *Building with Nature* concept into daily operation was at first a somewhat alien and exotic activity within a highly standardised environment. Size-wise, its initial contribution felt small and insignificant, a feeling strengthened by the disorienting experience of drifting on turbulent waters and surrounded by a dauntingly appearing dredging fleet. Despite the methodological greenness of this piloting exercise, which made daily operations susceptible to missteps, the pressure to perform was high. Failure costs due to potentially unintended environmental pollution or false-alarmed dredging operation halts could endanger project viability. This tension – between the solid ambition to conserve the marine environment and underlying economic drivers – made clear that a robust monitoring approach was essential.

After an initial period of 'learning-by-doing', broken-down and lost monitoring equipment, uncertain data-collection approaches, and misaligned communication channels with other departments involved in this vast operation, slowly but determined *Building with Nature* found its shape. Considerable investments were made to create a reliable and legitimate control system. Protocols were developed regarding measurement and analysis, as drivers of the overall embedment of environmental monitoring into dredging activity. Such practice-oriented learning in unforeseen environments forged a better understanding of the operating context, and the concretisation and allocation of tasks between all actors involved (Fig. 1).

Overall, *Building with Nature* turned out to be a laboratory for innovative environmental conservation approaches, of which

Fig. 1. Dredging Activity for Constructing the Khalifa Port and Industrial Zone in 2008, Abu Dhabi (Own Picture).

much – due to a lack of substantial reference methodology – had to be developed bottom-up. Despite continuous and evolving challenges of 24/7 dredging in one of the Middle East's largest Red Coral Reef areas (CORDIO, 2023), the project proved successful. Since then, other initiatives have followed suit (Boskalis, 2023; EcoShape, 2023). Globally, implications of *Building with Nature* to the marine conservation agenda seem obvious and a basis for research development to attain the UN SDG14 criteria.

LEGISLATION OF SUSTAINABILITY

Corporate Social Responsibility

The ocean and coasts are considered fundamental for creating collective wellbeing, inspiring cultural and recreational activities, providing essential ecosystem services, sourcing seafoods of high nutritional value, and promoting international trade and initiating sustainable economic growth (Wright et al., 2017). These considerations have partly influenced the initiation of the seventeen Sustainable Development Goals in 2015 and the specification of related targets and indicators (United Nations, 2015b, 2015a). The Khalifa Port and Industrial Zone project is a remarkable example of development where strict environmental requirements, despite the absence of globally binding legal forces, were requested and fulfilled accordingly. However, the dominant opinion remains that more pressing means are required to make sustainable economic behaviour a standard procedure. In this regard, sustainable development legislation is a long-debated topic.

An interesting example is a proposed Dutch law in 2022 on international corporate social responsibility. This proposition can be placed in a modern trend of governmental initiatives shifting towards more pressing instruments to combat climate change. A year earlier, during the Glasgow Declaration Summit in 2021, the Netherlands joined a group of various countries aiming to stop government support of foreign fossil fuel projects (Government.nl, 2021, 2022). The proposed law was supposed to anticipate European legislation and impose additional company obligations. Such obligations – *duty of care* – should prevent economic activity from negatively impacting sustainable development. Using legally binding responsibilities, *duty of care* could potentially induce the integration of various critical issues – like child labour, extortion, and modern slavery, as well as multiple forms of environmental pollution and nature destruction – within production, global business operation, and supply chain management (Winter et al., 2020). Additionally, larger companies could be forced to report on instances where principles of this law are accidentally or systematically violated, as well as force them to propose innovative solutions to limit or prevent such violations. Independent monitoring bodies would be able to sanction and support companies in the light of violations (NOS, 2023b).

Within *Building with Nature*, the social dimension also strongly influenced operation. Regardless of critical ongoing debates on immigration and labour (De Bel-Air, 2014), the multi-cultural involvement might arguably have produced much added value within the Khalifa Port and Industrial Zone project. An assertive, pragmatic, and hands-on attitude – a quality of work that might have become underrated in Western practice – often resulted in valuable solutions in the light of unexpected challenges. Also, on an interpersonal level, the culture-rich collaboration led to reappraised work-live relationships, a renewed prioritisation and interpretation of taken-for-granted norms, an acknowledgement that values are culturally embedded, and that positions and situations are relative of nature.

Nevertheless, despite the considerate benefits that social integration can bring to large construction projects, there are indications that such integration does not always manifest harmonically. One example includes a recent project in Manila Bay, where the construction of an airport complex was initiated within a biodiverse and fishing-culture-rich mangrove Forest area (NPO, 2023). In this regard, a critical discussion relates to how large-scale projects should be emotionally judged based on socio-cultural predisposed norms within a context of competitive global players. Arguably, this existing dilemma between global competition versus sustainable project development is a phenomenon that cannot be solved by placing responsibility on isolated actors but which needs to be drawn out broadly and systematically using a rational assessment and decision-making apparatus for supporting moral dilemmas (Stahl et al., 2020; Kim, 2022).

The tension between economic activity and social welfare is also visible underwater. Nature conservation initiatives often face a challenge: their potential impact on local employment and economic sectors (Jahanger et al., 2022). During studies conducted at Leiden University, within the context of a biodiversity monitoring project on Curacao for Naturalis in 2006, there was an apparent conflict between fishery and the preservation of the endangered queen conch *Strombus gigas*. Although the queen conch has historically been established as an important cultural symbol and source of income for local industry, scientists and managers in the Wider Caribbean region have long recognised the need for coordinated management. Despite numerous regional initiatives dedicated to sustainably balancing industrial activity and queen conch conservation, tangible results have been limited due to a lack of cooperative structures and political support (Prada et al., 2017).

It was during this research project that MatureDevelopment came into view, a non-profit organisation based in The Hague that generally aims to drive sustainable business development, mainly through the exploration of innovative technological approaches by re-imagining the intrapersonal, interpersonal and business dimensions of personal and business life.[2] As such, MatureDevelopment seeks to redirect the fundamental nature of established learning systems. It also actively engages in the mentoring of students from the further and higher education sectors to reinforce their skillsets and become immersed in the SDG missions.

The highly acclaimed Brazilian educator and philosopher Paulo Freire – generally considered a pivotal figure in the field of critical pedagogy – has strongly advocated the important reciprocity of the student-teacher relationship. Instead of being those who exclusively instruct, teachers are also taught in dialogue with students, resulting in both sides becoming jointly responsible for a process where all can grow (Freire, 1973). Freire's philosophy seems to provide an exciting perspective now relevant for the SDG14 remit.

He believed that the practice of education is seldom a neutral process since both teaching and learning can be considered sub-acts within larger political agendas. In such context, teaching aims to facilitate the integration of generations into the logic of present systems, potentially moderated by existing political notions of those who teach, consciously or subconsciously. Instead, he argued to encourage a 'practice of freedom' to create a learning environment that deals critically with reality and discover options to participate in the transformation of personal worlds (Mayo, 1999).

In this regard, the operation of MatureDevelopment seems to resonate strongly with Freire's views. During many of its projects over the past years, ranging from producing water quality and aquatic seafood systems to standardising biofuels and assessing bio-stimulants, finding fruitful solutions sprouted from a basis of conversation, discussion, reflection, and integration. The typical intergenerational approach built-in MatureDevelopment's mindset – through connecting the experience and expertise of industries with the novel ideas of the next generation in higher education, such as undergraduates and postgraduates as well as researchers – strongly follows this notion of transformative action and has proven to be enriching and advancing lifetime experiences of students and young stakeholders in academia and commerce.

Challenges of Sustainability Legislation

Although a law on corporate social responsibility could be a valuable instrument for realising sustainable projects, there is much debate regarding the effectiveness of such legislation (Ahlström, 2019). Supporters of sustainability legislation argue that most of its essence has long been legally required and, therefore, would only increase transparency on corporate social responsibility *de facto*. Also, it is believed that such a law could drive companies to connect their expertise with goals globally shared for climate, equality, and biodiversity (Stibbe, 2023). However, opponents argue that creating legislation on something vast and largely undefined can dramatically affect operations. When guidelines are transformed into legal duties, executives can be judicially reviewed and held personally liable. Due to the current mainly directive nature of the proposed law, its enactment could create a relatively large space for judiciary interpretation and potentially lead to lottery-like practices (NOS, 2023b).

In addition, it is argued that current forms of sustainability legislation not only concern the conditions under which projects are implemented – like human rights protection and the provision of healthy working conditions – but also the long-term impacts that the implementation of a project may create. This impact dimension concerns significant infrastructural and technological investments operating companies need to make, particularly about the global challenge of climate change. Since many investments have an expected lifetime of decades, opponents claim that the current unavailability of competitive technology for reaching long-term goals will influence the sustainability performance of organisations that must make such investments now (FD, 2023).

As a consequence of the increased operational ambiguity, more challenging and complex investment decisions, as well as the potential loss of competitive advantage as a result of sustainability legislation, there is increased fear that influential organisations might increasingly leave the public space of trading platforms, which would weaken incentives to report on corporate sustainability. Also, various companies have already stated that they will potentially relocate when stricter sustainability criteria are imposed, which could hurt national economies (NOS, 2023a). Therefore, it seems imperative that examples of legislative dialogue are also incorporated into teaching and research for students practicing the biosciences of SDG14, as it will enable a more secure basis for their application and dissemination of knowledge suited to the political and societal mandate for attaining genuine sustainability.

REPRESENTATION OF SUSTAINABLE DEVELOPMENT

Financial and Temporal Challenges

Wanting to be sustainable is often not the primary challenge; it is the road towards attaining this objective. Not only are companies struggling to address sustainable development in a comprehensive and economically viable manner, but it is a phenomenon visible across society. Although political efforts have often aimed to shape cultural dynamics and stimulate social innovation, adopting a sustainable lifestyle is arguably still a niche behaviour (Mont et al., 2014; Kashima, 2020). However, recent events in the energy market have shown that economic incentives to behave sustainably can work (Renewable.news, 2022; Rosenbloom et al., 2020; Boudet, 2019).

This nature of voluntariness also seems to slow down the realisation of the SDGs. Although this collection of objectives and associated indicators are widely used within operational contexts as blueprints for sustainable development, the lack of considered social, economic, and environmental trade-offs within SDG design and interpretation reduces the power to act and enforce (McCloskey, 2015; Schleicher et al., 2018; Kotzé et al., 2022). It might be unjust to blame the problematic translation of embracing SDGs into actual sustainable realisations as being merely instrumental for greenwashing or warming public opinion. This must be realised appropriately for the SDG14 mandate concerning Life Below Water for a credible and positive outcome. Future students within the higher education sector must be imparted with such fiscal and economic awareness through training and experience linked to their curriculums. This would further serve the objectives of SGD14 to reinforce its successful implementation.

In 2008, when the progressive nature of the UAE's development was highlighted by building Masdar, the world's first carbon-neutral city, all eyes were centred on how revolutionary concepts for sustainable cities could become realised (CBC News, 2008). After a somewhat slow start – providing room for scepticism regarding the overall intentions of this massive project – the making of Masdar has seemingly accelerated from visionary towards reality (Masdar City, 2023). Masdar City is exemplary for past projects that have shown that the realisation of sustainable development generally requires symbiotic structures, based on different people pursuing a common future, relationships that are full of complexity and challenges that need to be solved. Sustainable development is a form of development that comes slowly. It requires patience, a challenging condition in times where action is fast, where the dangers of changing climates and societies loom over the horizon of daily life.

The Interdependence of Economic, Political, and Societal Forces

Instead of rapid, radical behavioural changes, taking small steps towards sustainable development might be more fruitful. To embed sustainable development in our social DNA will be resource- and time-intensive. To succeed in a society that grows diverse and restless will demand effort and compromises from all involved (Cai et al., 2020).

The disappearance of leading companies could implicate a significant loss of expertise and labour, elements that are strongly required for sustainable practice. Responsible business, indeed, should be justified by the measures of society. It is finding a middle ground to polder towards results between the activist progressive and the laissez-faire conservative. Not striving for complete satisfaction but going for systematic optimisation.

As a result of the dichotomy between politics and corporations, the interpretation of sustainability often comes at the expense of harmony between state and economy. For example, legislation of sustainable development seems a double-edged sword. Is it a necessary stimulus for the industry to exchange current business models for ones less centred around fossil resources? Is this how governance can be redirected, by shifting towards initiatives that breathe sustainability throughout? Or does such policy only increase the dependency of nations that forerun sustainable development on others being less progressive?

A system based on voluntary behaviour will behave much differently compared to one in which sustainability is anchored in legislation. Companies can now follow existing guidelines within their context of operation, generally to an extent that does not jeopardise existing business models. Although such a voluntary approach offers potential escape modules to uphold the *status quo*, it also provides the flexibility to implement policies based on the interpretation of characteristic rapidly changing competitive environments. These features must also be reflected in teaching and learning commitments, including high school, further and higher education sectors. They are invaluable principles supporting innovation, enterprise, and the governance and implementation of the SDG14 framework. These core issues must be conveyed to future generations of academics and their students at all levels within the institutional framework; this should occur within a multidisciplinary strategy embracing the life sciences – as well as engineering, political and societal knowledge platforms – to be lenses for the focal awareness of future generations. Such a context will auger well for the SDG14 Life Below Water objectives as a timely example.

CONCLUSION – BUILDING WITH PEOPLE

Sustainable development contains a paradox. On the one hand, to sustain ecosystems, we need to be conservative. On the other hand, development is a core force that makes things move and progress, essential to the nature of being. Since its origin of existence, the natural world has always been in a dynamic state, regardless of human interference. The 'nature of nature' has always been to adhere to the Darwinian laws of adaptation. Although it might make sense to initially rule out human impact on nature conservation, it might also fall short of its aims. It is unlikely that any form of natural isolation is practically achievable. History has shown that restrictions on development do not necessarily work. First, enormous effort is often required to monitor, regulate, and effectuate bans. Second, creating a legitimate basis for imposing bans requires a long time, in which different stakeholders need to find a commonly agreed solution. Such delays can cause lingering sustainability initiatives to become dormant and rising movements to go static. Instead, one could try to integrate human behaviour within ecosystems sustainably. For such an approach to succeed, it must be clear and convincing for all involved stakeholders how mutually beneficial situations can be made in which nature and people can benefit from symbiotic relationships. In this regard, current research on innovation systems has shown that the inclusive integration of stakeholders is crucial for success (Lindner et al., 2016). However, developed models also highlight that change comes slowly, often because of conflict between short-term interests and long-term ambitions.

Nevertheless, the implementation of sustainable development seems to have evolved over time; it has been and continues to be, a continuous learning process; one that also seems increasingly incorporated into the educational system in curriculums and research activities. Various universities and colleges associated with the aquatic and marine areas are now practicing such topics and highlighting longer term SDGs such as SDG14 as key elements in their pedagogic planning.

This learning process can be observed at various levels of change: Entrepreneurs landing innovative ideas through crowdsourcing, large multinational corporations increasingly anticipating the adoption of emerging technology, university curricula with greater attention to interdisciplinarity, and a change in educational focus from fact-learning to problem-solving. The case study of *Building with Nature* shows that a more systematic embedding of nature and people in development projects can – although not unchallenged – create a potentially significant added value. Over the past years, the capital of the United Arab Emirates has radically transformed into a central power and has positioned itself on the radar of global initiatives (De Jong et al., 2019). Retrospectively, the proactive approach towards environmental protection can arguably be labelled as historically unconventional and progressively modern. When formalised as crucial success criteria, marine ecosystem conservation can integrate well within strategies based on open-mindedness, reared towards solving operational challenges, and aligned with goals set for sustainable development. The localised approach followed within the Abu Dhabi project, which is also strongly present within the scope of SDG14, highlights that the integration of regional multi-level perspectives in addressing life below water is increasingly being acknowledged (Wright et al., 2017).

Although scientific and technological progress once were seen as the main drivers of change – a belief that has centralised R&D in national and international policymaking – this perspective has broadened over time to a more inclusive one. Instead of underestimating the human dimension, new approaches that acknowledge the complex dynamics of our social systems, arenas populated by interacting innovative actors with competing interests, are maturing. Also, in relation to the initiatives explored within the context of MatureDevelopment, it has become evident that – through matching the visions of learning platforms and higher education with the voices of industry – intergenerational knowledge exchange can unlock new forms of learning and innovation for addressing many of modern world's challenges. In line with Freire's teachings regarding the value of the reciprocal teacher–student relationship, and through appreciating the value of dealing openly and critically with reality, options can be discovered to participate together in societal transformations and drive sustainable development.

Together, we can realise the embodiment of SDGs into the fabric of society, stimulated by the efforts of higher education, government, industry, and beyond. It has been shown that governance and societal integration towards supporting biodiversity and marine environmental stewardship can be applied to various scenarios of achieving sustainability. It can also underpin student comprehension of how SDG14 remit of Life Below Water can be effectively and sensitively realised whilst securing wealth and employment security in balance with nature. Having put *Building with Nature* to the test, it may be time to take the next step: An evolution of the human perspective, from nature-centric to all-encompassing. It might be time for *Building with People*.

NOTES

1. See also https://www.ecoshape.org/en/about/our-network/ [Accessed 31 May 2024].

2. See also https://maturedevelopment.com/ [Accessed 31 May 2024].

REFERENCES

Ahlström, H. 2019. Policy hotspots for sustainability: changes in the EU regulation of sustainable business and finance, *Sustainability*, 11(2), 499. doi: 10.3390/SU11020499

Boskalis. 2012. *Khalifa Port, Abu Dhabi*. Available at: https://boskalis.com/sustainability/cases/khalifa-port-abu-dhabi [Accessed 31 May 2024].

Boskalis. 2023. *Building with Nature*. Available at: https://boskalis.com/about-us/company-profile/building-with-nature [Accessed 31 May 2024].

Boudet, H.S. 2019. Public perceptions of and responses to new energy technologies, *Nature Energy*, 4(6), 446–455. doi: 10.1038/s41560-019-0399-x

Cai, M., Murtazashvili, I., Murtazashvili, J. B. and Salahodjaev, R. 2020. Patience and climate change mitigation: global evidence, *Environmental Research*, 186, 109552. doi: 10.1016/J.ENVRES.2020.109552

CBC News. 2008. *Abu Dhabi Unveils Carbon-Neutral City*. Available at: https://www.cbc.ca/news/science/abu-dhabi-unveils-carbon-neutral-city-1.747858 [Accessed 31 May 2024].

CORDIO. 2023. *Coral Species and Coral Reef Ecosystems Red Listing, Coastal Oceans Research and Development – Indian Ocean*. Available at: https://cordioea.net/ [Accessed 31 May 2024].

De Bel-Air, F. 2014. *Demography, Migration, and Labour Market in Qatar – Explanatory Note No. 8 Gulf Labour Markets and Migration (GLMM)*. Jeddah, San Domenico di Fiesole. Available at: https://cadmus.eui.eu/handle/1814/32431 [Accessed 31 May 2024].

De Jong, M., Hoppe, T. and Noori, N. 2019. City branding, sustainable urban development and the rentier state. How do Qatar, Abu Dhabi and Dubai present themselves in the age of post oil and global warming? *Energies*, 12(9), 26 pp. doi: 10.3390/EN12091657

de Vriend, H.J., van Koningsveld, M., Aarninkhof, S. G. J., de Vries, M. B. and Baptist, M. J. 2015. Sustainable hydraulic engineering through building with nature, *Journal of Hydro-environment Research*, 9(2), 159–171. doi: 10.1016/J.JHER.2014.06.004

EcoShape. 2023. *Pilots Archive*. Available at: https://www.ecoshape.org/en/pilots/ [Accessed 31 May 2024].

FD. 2023. *Boskalis-topman: 'Aanzien van Nederland op het wereldtoneel bladdert af'*. Available at: https://fd.nl/bedrijfsleven/1463506/boskalis-topman-aanzien-van-nederland-op-het-wereldtoneel-bladdert-af-nfa3caRLPuBA [Accessed 31 May 2024].

Freire, P. 1973. *Education for Critical Consciousness*, London, Continuum.

Government.nl. 2021. *The Netherlands Submits Adaptation Communication Ahead of COP26*. Available at: https://www.government.nl/documents/reports/2021/10/01/the-netherlands-submits-adaptation-communication-ahead-of-cop26 [Accessed 31 May 2024].

Government.nl. 2022. *The Netherlands Takes a New Step in Greening Export Credit Insurances*. Available at: https://www.government.nl/latest/news/2022/11/01/the-netherlands-takes-a-new-step-in-greening-export-credit-insurances [Accessed 31 May 2023].

Jahanger, A., Usman, M., Murshed, M., Mahmood, H. and Balsalobre-Lorente, D. 2022. The linkages between natural resources, human capital, globalization, economic growth, financial development, and ecological footprint: the moderating role of technological innovations, *Resources Policy*, 76, 102569. doi: 10.1016/J.RESOURPOL.2022.102569

Kashima, Y. 2020. Cultural dynamics for sustainability: how can humanity craft cultures of sustainability? *Current Directions in Psychological Science*, 29(6), 538–544. doi: 10.1177/0963721420949516

Kim, R.C. 2022. Rethinking corporate social responsibility under contemporary capitalism: five ways to reinvent CSR, *Business Ethics, the Environment & Responsibility*, 31(2), 346–362. doi: 10.1111/BEER.12414.

Kotzé, Louis J., Kim, Rakhyun E., Burdon, P., du Toit, L., Glass, L.-M., Kashwan, P., Liverman, D., Montesano, F. S., Rantala, S., Sénit, C.-A., Treyer, S. and Calzadilla, P. V. 2022. Planetary Integrity. In *The Political Impact of the Sustainable Development Goals*, Eds F. Biermann, T. Hickmann and C.-A. Sénit, pp. 140–171, Cambridge University Press. doi: 10.1017/9781009082945.007

Lindner, R., Kuhlmann, S., Randles, S., Bedsted, B., Gorgoni, G., Griessler, E., Loconto, A. and Mejlgaard, N. (Eds.) 2016. *Navigating Towards Shared Responsibility in Research and Innovation*, Eggenstein, Stober GmbH Druck und Verlag. Available at: www.res-agora.eu [Accessed 31 May 2024].

Mansourmoghaddam, M., Malamiri, H. R. G., Rousta, I., Olafsson, H. and Zhang, H. 2022. Assessment of Palm Jumeirah Island's construction effects on the surrounding water quality and surface temperatures during 2001–2020, *Water*, 14(4), 634. doi: 10.3390/W14040634

Masdar City. 2023. *Innovating Today for a Sustainable Future*. Available at: https://masdarcity.ae/en/discover/about-us [Accessed 31 May 2024].

Mayo, P. 1999. *Gramsci, Freire and Adult Education: Possibilities for Transformative Action*, London, Zed Books. doi: 10.1023/A:1017970330226

McCloskey, S. 2015. From MDGs to SDGs: we need a critical awakening to succeed, *Development Education Review*, (20). Available at: https://www.developmenteducationreview.com/issue/issue-20/mdgs-sdgs-we-need-critical-awakening-succeed [Accessed 31 May 2024].

Mont, O., Neuvonen, A. and Lähteenoja, S. 2014. Sustainable lifestyles 2050: stakeholder visions, emerging practices and future research, *Journal of Cleaner Production*, 63, 24–32. doi: 10.1016/J.JCLEPRO.2013.09.007

NOS. 2023a. *Boskalis dreigt met vertrek uit Nederland, 'nieuwe wet maakt ondernemen onzeker'*. Available at: https://nos.nl/artikel/2458887-boskalis-dreigt-met-vertrek-uit-nederland-nieuwe-wet-maakt-ondernemen-onzeker [Accessed 31 May 2024].

NOS. 2023b. *Waarom is er zoveel kritiek op een wet voor maatschappelijk verantwoord ondernemen?* Available at: https://nos.nl/nieuwsuur/artikel/2460266-waarom-is-er-zoveel-kritiek-op-een-wet-voor-maatschappelijk-verantwoord-ondernemen [Accessed 31 May 2024].

NPO. 2023. *Frontlinie - De Baggerbaai op de Filipijnen*, The Netherlands, VPRO. Available at: https://www.npostart.nl/frontlinie/09-02-2023/VPWON_1346244 [Accessed 31 May 2024].

Prada, M. C., Appeldoorn, R. S., van Eijs, S. and M., Manuel Pérez. 2017. *Regional Queen Conch Fisheries Management and Conservation Plan*, Technical Paper No. 610, Rome, FAO Fisheries and Aquaculture.

Renewable.news. 2022. *Heat Pumps in the Netherlands Sold Out Until the End of 2022*. Available at: https://www.renewable.news/climate-change/heat-pumps-in-the-netherlands-sold-out-until-the-end-of-2022/ [Accessed 31 May 2024].

Rogers, C.S. 1990. Responses of coral reefs and reef organisms to sedimentation, *Marine Ecology Progress Series*, 62, 185–202. Available at: https://www.int-res.com/articles/meps/62/m062p185.pdf?origin=publication_detail&sa=U&ei=jsohVITbOJXXoAS184DADQ&ved=0CEcQFjAJ&usg=AFQjCNE80wJFFt5dhebBkxkWlxL7r0hlrw [Accessed 31 May 2024].

Rosenbloom, D., Markard, J., Geels, F. W. and Fuenfschilling, L. 2020. Why carbon pricing is not sufficient to mitigate climate change – and how "sustainability transition policy" can help, *Proceedings of the National Academy of Sciences of the United States of America*, 117(16), 8664–8668. doi: 10.1073/PNAS.2004093117/ASSET/9016B670-537A-4644-A00C-4CEAB9D56EDD/ASSETS/PNAS.2004093117.FP.PNG

Schleicher, J., Schaafsma, M. and Vira, B. 2018. Will the Sustainable Development Goals address the links between poverty and the natural environment?, *Current Opinion in Environmental Sustainability*, 34, 43–47. doi: 10.1016/J.COSUST.2018.09.004

Stahl, G.K., Brewster, C. J., Collings, D. G. and Hajro, A. 2020. Enhancing the role of human resource management in corporate sustainability and social responsibility: A multi-stakeholder, multidimensional approach to HRM, *Human Resource Management Review*, 30(3), 100708. doi: 10.1016/J.HRMR.2019.100708

Stibbe. 2023. *(Internationaal) Maatschappelijk Verantwoord Ondernemen*. See also: https://www.stibbe.com/nl/publications-and-insights/actualiteiten-internationaal-maatschappelijk-verantwoord-ondernemen-2#:~:text=In%20februari%202023%20lieten%20de,de%20minister%20van%20Buitenlandse%20Handel [Accessed 31 May 2024].

The Brundtland Commission. 1987. *Our Common Future*.

Tuttle, L.J., Johnson, C., Kolinski, S., Minton, D. and Donahue, M. J. 2020. How does sediment exposure affect corals? A systematic review protocol, *Environmental Evidence*, 9(1), 1–7. doi: 10.1186/S13750-020-00200-0/TABLES/1

United Nations. 2015a. *Global Indicator Framework for the Sustainable Development Goals and Targets of the 2030 Agenda for Sustainable Development Goals and Targets (from the 2030 Agenda for Sustainable Development) Indicators.*

United Nations. 2015b. *Transforming Our World: The 2030 Agenda for Sustainable.* Available at: https://sdgs.un.org/2030agenda [Accessed 31 May 2024].

Winter, J. W., de Jongh, J. M., Hijink, J. B. S., Timmerman, L., van Solinge, G., Lennarts, M. L., Wezeman, J. B., Bulten, C. D. J., Bartman, S. M., Lokin, E. C. H. J., Wuisman, I. S., Vletter-Van Dort, H. M., Schwarz, C. A., Verbrugh, M. A., Roest, J., Raaijmakers, G. T. M. J., Koster, H., Kemp, B., Olaerts, M., Joosen, E. P. M., Boschma, H. E., Verdam, A. F., Schutte-Veenstra, J. N., Willems, J. H. M. and Rensen, G. C. J. 2020. Naar een zorgplicht voor bestuurders en commissarissen tot verantwoorde deelname aan het maatschappelijk verkeer, *Ondernemingsrecht*, 2020(7), 471–474. Available at: https://hdl.handle.net/1887/139223 [Accessed 31 May 2024].

Wright, G., Schmidt, S., Rochette, J., Shackeroff, J., Unger, S., Waweru, Y. and Müller, A. 2017. *Partnering for a Sustainable Ocean: The Role of Regional Ocean Governance in Implementing Sustainable Development Goal 14*. TMG. doi: 10.2312/iass.2017.011

10

KNOWLEDGE HIERARCHIES IN POLICY-LEVEL TRANSLATIONS OF SDG14: INSIGHTS FROM HUMAN–SALINITY RELATIONS IN SOUTH INDIA AND VIETNAM

Richard Pompoes

Wageningen University and Research, Netherlands

ABSTRACT

The study underlying this chapter investigates how diverse actors in the Cauvery Delta, India, and the Mekong Delta, Vietnam, understand and live with water salinity. In focusing empirically on river deltas, this chapter addresses some of the SDG14 targets, as SDG14.2 ('Protect and restore ecosystems') and 14.5 ('Conserve coastal and marine areas') refer to the sustainable management of coastal areas as crucial targets for SDG14. Based on interviews with land users in the two deltas, in tandem with analyses of salinity maps and other policy-level knowledge artefacts, this chapter shows how, in some cases, only particular forms of knowledge are represented at the policy level, while many of the diverse viewpoints of land users are rendered invisible. In this way, delta management only meets the concerns of a select few, often professional elites, and limits land users from taking ownership of their own realities. This chapter concludes with the recommendation for water professionals, scholars, and practitioners alike, to be more open-minded, modest, and attentive to difference, by engaging more seriously with interdisciplinarity and cultivating sensibilities for listening to 'smaller' water stories.

Keywords: Cauvery Delta; Mekong Delta; salinity maps; knowledge hierarchies; SDG14; water professionals

INTRODUCTION

Sustainable Development Goal 14 'Life Below Water' recognises the need for action to achieve a sustainable future for our oceans and coastal ecosystems. Policy, science, and water management practices globally show efforts in working on the ocean- and coastal-related sustainability issues (WWF, 2020). While SDG14 is commonly referred to as the sustainable management of the ocean, SDG14.2 ('Protect and restore ecosystems') and 14.5 ('Conserve coastal and marine areas') also explicitly refer to the sustainable management of coastal areas as crucial targets for SDG14. Coastal zones can include sandy and rocky shores, mangroves, estuaries, and wetlands, among others. While the significance of transitioning towards sustainable oceans and coastal areas is widely recognised, Liebrand (2019), among others, contends that in policymaking, policy actors and other professionals at higher levels of governance may strategically overlook insights from qualitative social sciences. This insinuates that to address SDG14, policy actors might strategically translate only particular types of knowledge into policy. Nugroho et al. (2018) concur that in policymaking, professionals at higher levels of governance often prioritise scientific knowledge, which is often thought of as natural scientific knowledge, over so-called local knowledges and the less 'exact' forms of knowledge. This can be problematic as coastal zones serve essential ecological, economic, social, and cultural functions in diverse coastal communities that rely on these spaces for their livelihood activities, religious celebrations, recreation, and more (Rao and Majumdar, n.d.).

Research suggests that current efforts to achieve SDG14 and its targets are not meeting expectations (WWF, 2020; Dauvergne, 2018). This chapter, therefore, explores how existing knowledge hierarchies in coastal zone management affect efforts of achieving SDG14 and proposes further actions for SDG14 to challenge existing knowledge hierarchies, to contribute towards fairer governance of coastal zones. The Cauvery Delta, India, and the Vietnamese Mekong Delta are used as case studies to show how sustainability efforts are translated into policy in the two coastal regions. By unpacking the two cases where policy-level artefacts are strategically overlooking social science insights and favour natural scientific knowledge, this chapter intends to contribute to the body of work that problematises knowledge hierarchies in natural resource management.

Parts of this chapter have been published in and adapted from the author's master's thesis (see Pompoes, 2022).

FURTHER ACTIONS FOR SDG14

Based on the underlying master's thesis project of this chapter, viewpoints on further actions for SDG14 emerge. The recommended further actions for SDG14 are intimately linked to further actions for the higher education of water professionals. The recommendations can be summarised in two key points:

1. **Promote and help cultivate an attentiveness to difference during the higher education of future water professionals, by engaging more seriously with interdisciplinarity.** It is valuable to (learn to) describe the lived experiences of those most vulnerable to the effects of climate change by listening attentively and observing practices, also in coastal and delta areas. Disciplines outside the domain of the natural sciences can provide methodological insights on how to engage more meaningfully with diverse actors. This, too, requires acknowledging that achieving SDG14 is always an inherently social and political undertaking.

2. **Encourage more symmetrical analyses of sustainability problems,** i.e., by looking at various levels of governance and contrasting the viewpoints of farmers with policy or by grounding hydrological analyses with ethnographic insights. Also, encourage more symmetrical conversations between actors at different levels of governance.

The Master's Thesis Project

Coastal areas and intertidal zones are often characterised by brackish water environments, particularly river deltas. Deltas are some of the most populous regions in the world and simultaneously also some of the most climate-vulnerable regions. Saline ingress, a significant sustainability concern in many deltaic coastal zones, occurs naturally but is exacerbated by anthropogenic activities and is expected to increase significantly with sea-level rise and land subsidence (cf. Eslami et al., 2021). Communities in Asian river deltas increasingly shift to brackish water shrimp aquaculture as a commercial practice, when water and soil salinisation increasingly challenge existing agriculture practices. Aquaculture promises higher returns than conventional agriculture, and its adoption is often promoted by local governments and foreign investors, which link an adoption of brackish water aquaculture with an increase in climate adaptation and reduction in climate vulnerabilities; this is often associated with seawater ingress and salinisation of soil and groundwater (Paprocki, 2018; Pratheepa et al., 2022).

Brackish water shrimp aquaculture is already a relatively common commercial practice in the coastal stretches of the Cauvery Delta in Tamil Nadu, South India, and in the Vietnamese Mekong Delta. The Tamil Nadu Government views aquaculture as an important industry, in line with initiatives of the Central Government of India that has major plans for aquaculture development. The Mekong Delta is of key significance to the Vietnamese economy, with shrimp aquaculture being an important sector since the 1990s contributing to the Vietnamese aquaculture export economy. Pangasius and snakehead fish have a long history in fisheries in Vietnam and are now also increasingly being cultured in the Mekong Delta, constituting key aquaculture practices in the region. More recently, however, there are emergent sustainability concerns associated with brackish water shrimp aquaculture in the two deltas, many of which revolve around salinity. Saline water from brackish water shrimp farms can leach into adjacent agriculture fields, and groundwater over-extraction can increase seawater ingress, exacerbating climate change effects and increasing climate vulnerabilities of coastal communities (Paprocki, 2018; Pratheepa et al., 2022). There are existing policy efforts addressing sustainability concerns surrounding aquaculture and salinity in the two deltas, including master plans and stricter shrimp farm licencing policy.

In light of the sustainability concerns revolving around brackish water shrimp aquaculture, the objective of the master's thesis project was to unpack how diverse actors understand and frame salinity and sustainability concerns associated with aquaculture in the coastal zones of the two deltas, by learning from ground-level realities and understanding how professionals at the policy level translate these ground-level realities into policy and planning. This involved data collection through unstructured interviews with farmers, scientists, and policy actors. Altogether, the master's thesis research involved analysing 114 interviews, of which about half were conducted by co-researchers at an Indian and a Vietnamese research institute (Pompoes, 2022). The interviews were intended to explore how actors at various levels of governance understand and frame salinity, where framing can be understood as actors highlighting different aspects of a situation or phenomenon as problematic, urgent, or relevant. In addition, policy-level documents, media, and salinity maps were analysed to uncover how aquaculture and delta management policies and projects manifest in the two research sites.

The approach of studying the diverse perspectives at various levels of governance in a symmetric way, and encountering them by being, inspired by Latour (2005), only developed over time. The initial plan was to describe how communities of water professionals negotiate diverse interests in the two deltas and how this shapes aquaculture development and delta management. As data collection and analysis progressed, discrepancies between the viewpoints of actors at the ground level and policy, so of farmers and policy, became increasingly apparent. This dissonance became particularly evident when conducting a preliminary literature review of the historical developments of aquaculture in the Cauvery Delta and along the Coromandel Coast more widely. Conflicts between fisherfolk and resourceful multinationals characterised the 1990s along the South Indian coastline. To many, these conflicts were the first indication of how brackish water shrimp aquaculture development can foster a culture of encroachment by resourceful investors and upper castes (cf. Adduci, 2009). The dissonances provoked by the expansion of aquaculture and resistance movements in the Cauvery Delta called for an approach in which the viewpoints of actors at various levels of governance are given serious critical attention. To describe how knowledge is negotiated by aquaculture professionals, their viewpoints were contrasted with those of actors at lower levels of governance, in symmetry. Only looking at one level of governance would not have uncovered the controversies surrounding the management and development of coastal zones.

Diverse Understandings of Salinity in Coastal Zones

Case 1. Understandings and Framings of Salinity by Agriculture Farmers in Nagapattinam, Tamil Nadu

As data collection and analysis progressed, it became increasingly apparent that there are diverse ways coastal communities in the two Asian deltas live with and use the coastal zones. In the coastal district of Nagapattinam in the Kaveri Delta, for example, cropping patterns change over time in response to river discharge, rainfall patterns, monsoon seasons, and salt intrusion. This diversity is illustrated in Fig. 1.

The thesis focused on how various communities understand, frame, and respond to salinity. This figure highlights that there are diverse understandings and ways of how communities make use of salinity in the Kaveri Delta. For example, salt intrusion during the Southwestern Monsoon (SW) might be favourable for a shrimp farmer just starting their second cropping cycle of the year. In contrast, a rice farmer must adapt their cultivation to the drop in freshwater availability by cultivating plants producing seeds and

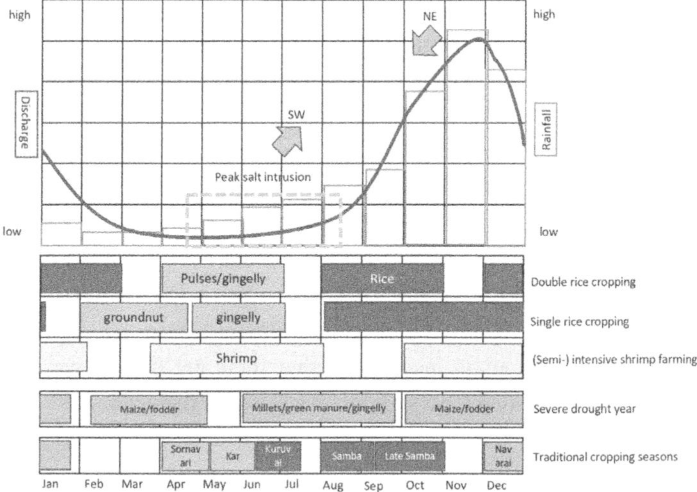

Note: River discharge, rainfall, and crop cycle lengths are not true to scale but only provide an approximate indication. *Source*: Adapted from Pompoes (2022).

Fig. 1. Illustrative Representation of Cropping Patterns, Rainfall, and River Discharge in the Nagapattinam District of the Kaveri Delta.

nuts, such as gingelly, groundnut, or pulses. This figure is, therefore, illustrative of the diversity of ways humans live with water in the Kaveri Delta, suggesting that water availability or salinity at a given time might have different meanings for different actors. This is further emphasised by the traditional cropping seasons employed in the delta, which suggests that farmers have their own ways of making sense of and understanding cropping seasons, which might be different to how cropping seasons are understood at a policy level in India. Water realities are, indeed, diverse in the delta – also when considering that this figure is only applicable to the district of Nagapattinam, one of the four coastal districts of the Kaveri Delta.

The diversity of water realities is also highlighted by how various actors at the ground level, primarily farmers, speak of salinity when asked about it in interviews. Some interviewed agriculture farmers from Vaimedu, Nagapattinam, voiced concerns over the over-extraction of groundwater by shrimp farms causing seepage of saline seawater into freshwater aquifers. The farmers reported being reliant on these aquifers, and one farmer reported seawater intrusion induced by shrimp farm borewells as the biggest challenge in his personal and business life. On the other hand, multiple interviewed aquaculture farmers in the same village reported that they are not concerned about salinity, not regarding their farm or household needs, nor regarding adjacent agriculture fields. This seems to suggest that high salinity concentrations are more worrisome for the interviewed agriculture farmers than for shrimp farmers in Vaimedu, even though brackish water shrimp farms also rely on maintaining particular salinity levels. The case of Vaimedu is just one village in the Kaveri Delta, again insinuating that there are diverse ways in which humans understand salinity and water in the delta.

Case 2. Understandings and Framings of Salinity by Rice-Shrimp Farmers in the Mekong Delta

Also in the Mekong Delta, farmers understand and live with salinity in diverse ways, as shown in Fig. 2.

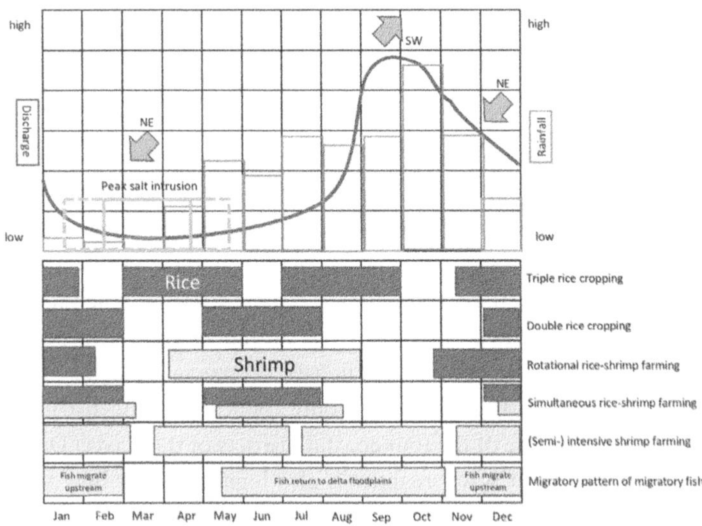

Note: River discharge, rainfall, and crop cycle lengths are not true to scale but only provide an approximate indication. *Source*: Adapted from Pompoes (2022).

Fig. 2. Illustrative Representation of the Mekong Delta Cropping Patterns, Rainfall, and River Discharge.

Fig. 2 highlights some of the land uses in the Mekong Delta – a site of diverse land uses and livelihoods. In addition to various rice and shrimp cropping modes, Fig. 2 also illustrates the migratory pattern of migratory fish in the Mekong Delta, which travel up- or downstream depending on the season and extent of saline intrusion (MRC 2002, as cited in Pompoes, 2022). The migration patterns of fish shape the livelihood activities of fisherfolk fishing along the Mekong River and its distributaries. At the moment in time when fisherfolk or rice farmers must adjust their livelihood activities in periods of drought and peak saline ingress, brackish water shrimp farmers start another shrimp cycle. However, shrimp farmers in the Mekong Delta also have diverse ways of understanding and using water and salinity. For example, one rice-shrimp farmer reported a farmer-led initiative in 2001 in which sluice gates in Bac Lieu province, near the coast, were destroyed to let saline water enter their fields to enable brackish water aquaculture. This farmer viewed the destruction of the sluice gates as having brought prosperity to his community. This was at a time when conventional rice farming increasingly became unviable due to low economic returns and high salinity levels in the coastal zones. During the interview, the farmer shared that the destruction of the Lang Tram salinity control sluice gates on the Ca Mau peninsula, which resulted in saline water entering his plot, was a turning point in his life, and that because of salinity, '[...] *we are happy now*'. (Pompoes, 2022 p.72). Other shrimp farmers from the delta spoke in more nuanced ways about salinity. A rice-shrimp farmer located in Tra Vinh province, about 20 km upstream from the coast, reported that '*salinity and water contamination are challenging our water security*' (p. 73) and that because of salinity, they now had to collect rainwater for domestic uses. The accounts of these two farmers suggest that water security can mean different things for different people, as the rice-shrimp farmers from Bac Lieu Province who partook in the removal of sluice gates in 2001 might perceive water security as brackish or saltwater security, while the rice-shrimp farmer in Tra Vinh Province makes sense of water security as freshwater security. Another shrimp farmer from Ca Mau Province observed high salinity levels during the dry season by looking at the water at night. He described seeing that salinity levels are high when '*the water is lightening*' (p. 75) at night.

The accounts of the agriculture and rice-shrimp farmers from the Cauvery and Mekong Delta illustrate that agriculture and aquaculture rely on particular, yet diverse water salinity levels. The ways in which actors understood salinity and water security differ drastically. Shrimp farmers with intensive, closed systems often thought of salinity in terms of pond salinity, while nature-based rice-shrimp farmers or agriculture farmers often thought of salinity as salt *intrusion*. During the interviews, it became apparent that land users had diverse ways of observing, measuring, and managing salinity, for example, by looking at the water 'lightening' at night, suggesting that actors can have their own understandings of the phenomenon that scientists call 'salinity'.

The diverse ways in which actors at the ground level, primarily farmers, understand and live with salinity and water in the two deltas suggest that policy actors have to translate, or somehow navigate, this diversity to produce policy and projects. To determine how policy actors translate this diversity, a symmetrical analysis was employed, by contrasting the accounts of actors at the ground level with accounts of policy actors, policy documents, and other knowledge artifacts of policy actors.

Salinity Management in the Cauvery and Mekong Deltas

While the ways in which farmers and other land users in the Cauvery and Mekong Deltas understand and live with salinity and water are diverse, policy on delta and coastal management is reductive and does not sufficiently reflect this diversity. The manifestations of policy and the accounts of policy actors differed between the two deltas, but there was regularity in how actors at the policy level strategically translated particular only particular understandings of salinity and water into policy, often to legitimise and facilitating the building of hydraulic infrastructure. The thesis research unpacked how this specific understanding and management of coastal zones can manifest in hydraulic structures, science, and media, of which the former two constitute cases unpacked in the subsequent sections.

Case 3. Seawater Intrusion Map of the Cauvery Delta

The map in Fig. 3 illustrates how cartographers can mobilise particular forms of knowledge to strengthen the voice of the government and water resources engineers who seek to control the freshwater zone in the Cauvery Delta through hydraulic infrastructure. The map in Fig. 3 is retrieved from the 2011 Asian Development Bank (ADB) report titled '*Support for the National Action Plan on Climate Change Support to the National Water Mission*', in which a strategic adaptation framework is proposed for the Cauvery Delta. This adaptation framework culminated in an adaptation project of a subbasin of the delta, initiated in 2016, involving irrigation channel improvements, building flood control structures, pump stations and tail-end regulators (ADB, n.d., as cited in Pompoes, 2022).

The map in Fig. 3 frames salinity as one-dimensional, unseasonal, and as the intrusion of seawater. The map nor the report in which the map is published does not specify whether the map refers to peak seawater intrusion levels or to which season these seawater intrusion lines are applicable. Instead, this map portrays salinity as a line on a map, reducing all associated variables into two tangible salinity lines. This removes any potential uncertainty, as a viewer of this map is made to imagine a tangible salinity that is happening irrespective

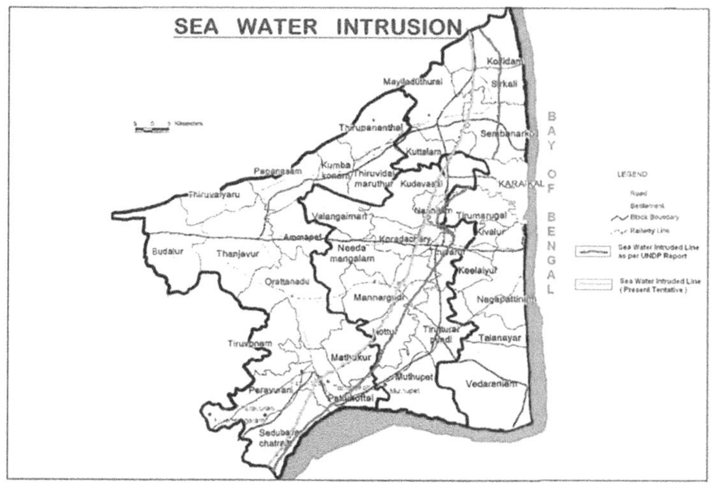

Source: ADB (2014).

Fig. 3. Seawater Intrusion Map of the Kaveri Delta.

of the season or other potential variables. The ADB report describes this map as '*showing the freshwater/saline boundary*' of the delta, which implies that salinity is reduced to a dualism of saline/seawater versus freshwater. Deltas are characterised by their fluidity and intermediary nature, as intertidal meeting places between sea and land, as argued by Morita and Jensen (2017), which is obscured by the static rendering of the delta and salinity in this map.

Framing salinity, or seawater intrusion, at the scale of the delta, is a political choice, as it frames the Cauvery Delta as an administrative and homogenous unit, despite the delta being diverse and not having regional governance bodies or administrative boundaries. However, framing the scale as a delta on this map makes the viewer imagine it as a tangible entity, thus effectively rendering the delta as a state space under the government's control and as ready for development. The delta on this map is also framed as a freshwater watershed, with seawater constituting its boundary, suggesting that salinity is a defining characteristic of the delta's boundary. This implies that when seawater intrudes, the size of the delta effectively decreases, rendering smaller the space for action, which in turn mobilises a powerful narrative in which dealing with seawater intrusion is an emergency, requiring extraordinary state intervention. The language in this map also connotes that seawater intrusion is happening *to* the delta, as the terms '*seawater*' and '*intrusion*' connote an intrusive movement of seawater into the delta system, an encroachment.

This map's intention is clear; salinity is framed as a threat linked to potential sea-level rise, as a phenomenon happening to the delta, by producing a 'correct' model of the delta and seawater intrusion. All of this is concealed behind a seemingly objective science, in which the diverse social realities are hidden behind a blank, homogenising representation of the region. In reducing salinity to a line, or two lines, this map produces an authoritative image that does not require explanation. This map, therefore, reinforces the voices of policy actors, water resource engineers, and government actors seeking to control the Cauvery Delta freshwater zone with hydraulic infrastructure. In doing so, this map gives a voice to these particular actors, who are professionals and not actors at the ground level, rendering the delta a space ready to be developed by these professionals.

Case 4. TV Broadcast on the 2020 Salinity Intrusion in the Mekong Delta

The thesis research also uncovered similar policy-level framings of salinity in the Mekong Delta, as shown in Fig. 4. This figure illustrates how the Vietnamese government can mobilise particular forms of knowledge through state media to strengthen its voice and legitimise infrastructure initiatives that keep salinity at bay.

Based on a translation of the written and spoken language in the video, the film still in Fig. 4 shows a newscaster of Vietnamese state-owned broadcasting station VTV presenting a map of the Mekong Delta. The map includes the projected levels of salt intrusion for the dry season of 2020 and the past salt intrusion levels of 2015–2016, both of which were among the years with the highest peak salt intrusion levels recorded in the Mekong Delta. The box on the bottom right-hand side of the map reads '*75KM cannot get freshwater*', meaning that saltwater was projected to displace freshwater 75 km upstream from the coastline. Throughout the video, the newscaster reports that salt intrusion regularly implies significant economic losses for the region and asserts that the upcoming salt intrusion of 2020 would be harsher and more severe than Vietnam's colonial past and foreign invasions.

Source: Vân Phạm (2020).

Fig. 4. Film Still of a News Broadcast on VTV on Salt Intrusion in the Mekong Delta.

This video, therefore, reduces the complexity of living with salinity and the diverse meanings associated with it, to salinity being something entirely detrimental to all. The video also showcases measures of the government which are framed to already be successful at keeping salinity at bay. Salt intrusion in this TV broadcast is, therefore, framed as an external threat to the hydrological freshwater watershed, associated with external factors of seawater and sea level rise, and as a problem that can only be solved through state intervention. Like the salinity map of the Cauvery Delta in Fig. 3, the TV broadcast reduces the complexity of salinity to a singular salt intrusion line, rendering invisible the seasonality and temporality of salt intrusion. As the two maps suggest, salinity maps, and maps more broadly, tend to convey the image of being based on science and assume its authoritative stance. This stance is further employed in this broadcast as a segment showcases interviewed scientists explaining the seriousness of the salt intrusion. Portraying the Mekong Delta, the most significant region for agriculture in Vietnam, as threatened by salinity or salt intrusion mobilises a viewpoint among the viewers of the broadcast, in which extraordinary state intervention in the form of hydraulic structures is legitimised – even by comparing the problem of salt intrusion with Vietnam's colonial history and framing it as an external problem, a '*saline intrusion emergency*'.

Interviews with land users suggested that TV broadcasts such as the one on VTV are watched by residents of the Mekong Delta, as three interviewed farmers reported watching TV to learn about salinity levels around their farms. This hints at the potential impact of narratives such as the one promoted in this broadcast. When, therefore, evaluating the title of the TV broadcast, '*saline intrusion emergency*' and asking for whom saline intrusion might be an emergency, then it becomes apparent that it is an emergency for those who enact water security only in terms of freshwater security. This effectively obfuscates the viewpoints of those who require brackish or saline water, such as brackish water shrimp farmers. Rendering invisible those who rely on the presence of salinity legitimises state intervention in maintaining the freshwater hydrological watershed status quo.

DISCUSSION AND CONCLUSION

The previous sections set out to illustrate the diverse ways in which salinity is perceived and dealt with by farmers, as well as actors at higher governance levels in the Cauvery and Mekong Deltas. Policy-level initiatives, such as the salinity map and TV broadcast, are reductive of this multiplicity and do not reflect all these diverse viewpoints. This implies that the findings of the master's research also coincide with the assertions of Nugroho et al. (2018) and Liebrand (2019), among others, that professionals at higher levels of governance may strategically overlook insights from qualitative social sciences and prioritise natural scientific knowledge over less exact forms of knowledge. The analysis of the map also shows how seemingly objective science can be called into question.

Reducing knowledge multiplicities into one singular knowledge description is unavoidable in policymaking. Knowledge descriptions, also in other domains, are always reductive in this sense. However, here multiple understandings and ways of living with salinity are reduced into narratives that legitimise the viewpoints and actions of select professional elites and only a subset of land users. In this way, delta management does not meet the views of all, or at least not enough, of the actors living and working in the delta, many of whom are already in precarious socio-economic situations. Furthermore, 'natural scientific' knowledge, as also mobilised in the salinity map and TV broadcast, can yield valuable perspectives on managing coastal and ocean zones and contribute significantly toward achieving SDG14. It is problematic, however, if it is the only knowledge translated by policy actors, when they address concerns that are, too, inherently social. After all, social aspects need to be considered to achieve sustainable ocean and coastal zones, as argued by Haward and Haas (2021).

This reductive translation effectively reduces the positions of many others as invalid, which can have profound implications on their lived experiences. Coastal and ocean management can then thus render residents' objects of development and prevent them from setting their own agendas and taking ownership of their regions' development and their own realities. Therefore, this chapter and the thesis upon which it is based aim to underline that deltas and coastal zones are socio-spatial entities. Natural resource management also implies struggles over knowledges, meaning, norms, identity, authority, and narratives (Boelens et al., 2016, as cited in Pompoes, 2022). This also suggests that framing coastal zones, deltas, or even oceans as such carries the risk of producing a static framing, defined by borders drawn on a map, instead of showcasing these socio-spatial entities as constructed and political interfaces of actors, interests, and knowledges. This implies that referring to SDG14 as the goal addressing life below water, runs the risk of obfuscating the complex social realities that are part of deltaic, coastal, and oceanic socio-spatial systems. The results of the thesis showed how coastal zone management in the two deltas could make elite agendas practicable but rarely those of residents at lower levels of governance. Through coastal and delta management policies, which can be based on strategic translations of specific forms of knowledge, the coastal and delta landscapes are continuously reshaped while also shaping the residents' diverse realities.

A takeaway for higher education institutions that educate future water professionals can be to encourage more interdisciplinary curricula, so learning from the social sciences and humanities, from their ideas, concepts, and methods, for example, on how to do meaningful fieldwork or how to engage with the vulnerable. This, too, implies learning to listen to the stories of actors, even if these might seem small or insignificant. Furthermore, higher education curricula should encourage students to look at sustainability problems more symmetrically, for example, by engaging with both farmers and policymakers. Governance processes can, too, benefit from more symmetrical conversations between farmers and policymakers; a two-way dialogue to accept ground-level communities as bearers of knowledge and involved in shaping the water contexts. To challenge existing knowledge hierarchies, future professionals should try to embrace open-mindedness and interdisciplinarity, reflexivity, and modesty, so that it is not only particular forms of knowledge shaping ocean and coastal management.

A more serious emphasis in science and academic education on tending to ground-level realities of those reliant on coastal and ocean resources is crucial for sustainable development and achieving SDG14. This is also what this chapter aims to do – to inspire future generations of water professionals to look at the politics of water management and to listen. There is not only value in listening to the varied experiences of humans to develop solutions tailored to local circumstances, but also to make sure these varied experiences, understandings and ways of living in our water worlds do not disappear.

ACKNOWLEDGEMENTS

This chapter is largely based on the author's master's thesis research, which was supported under the 2018 Prince Claus Chair at Utrecht University, The Netherlands, involving researchers from Utrecht University (UU), Ashoka Trust for Research in Ecology and the Environment (ATREE) and IHE Delft Institute for Water Education. The co-researchers contributed significantly to the master's thesis research through regular discussions and support with collecting, analysing and translating interview data. The overall objective of the wider research project revolved around uncovering the modes, institutions and understandings of land and water uses in the Cauvery Delta.

REFERENCES

ADB. 2014. *India: Climate Adaptation in Vennar Subbasin in Cauvery Delta Project*. Available at: https://wrd.tn.gov.in/pdfs/Draft_Initial_Environmental_Examination.pdf

Adduci, F. 2009. Neoliberal Wave Rocks Chilika Lake, India: conflict over intensive aquaculture from a class perspective, *Journal of Agrarian Change*, 484–511.

Dauvergne, P. 2018. Why is the global governance of plastic failing the oceans? *Global Environmental Change*, 22–31.

Eslami, S., Hoekstra, P., Minderhoud, P., Trung, N., Hoch, J., Sutanudjaja, E., ... van der Vegt, M. 2021. Projections of salt intrusion in a mega-delta under climatic and anthropogenic stressors, *Communications Earth & Environment*.

Gardner, S. 2014. Bridging the divide: tensions between the biophysical and social sciences in an Interdisciplinary Sustainability Science Project, *Environment and Natural Resources Research*.

Haward, M. and Haas, B. 2021. The need for social considerations in SDG 14, *Frontiers in Marine Science*.

Klevan, T., Karlsson, B. and Grant, A. 2019. The color of water: an autoethnographically-inspired journey of my becoming a researcher, *The Qualitative Report*, 1242–1257.

Latour, B. 2005. *Reassembling the Social – An Introduction to Actor-Network-Theory*, Oxford University Press.

Liebrand, J. 2019. Viewpoint – the politics of research on farmer-managed irrigation systems in Asia: some reflections for Africa, *Water Alternatives*, 129–145.

Macfarlane, B. 2014. Dualisms in higher education: a critique of their influence and effect, *Higher Education Quarterly*.

Morita, A. and Jensen, C. 2017. Delta ontologies infrastructural transformations in the Chao Phraya Delta, Thailand, *Social Analysis*, 639–671.

Nugroho, K., Carden, F. and Antlov, H. 2018. *Local Knowledge Matters*. Policy Press.

Paprocki, K. 2018. Threatening dystopias: development and adaptation regimes in Bangladesh, *Annals of the American Association of Geographers*, 955–973.

Pompoes, R. 2022. *A Tale of Two Deltas – Unpacking the Aquaculture-Salinity Interface in the Kaveri and Mekong Delta*, Master's thesis, Utrecht University.

Pratheepa, C., Raj, R. and Sinha, S. 2022. The socio-ecological contradictions of land degradation and coastal agriculture in south India, *E Nature and Space*, 1–21.

Rao, A., & Majumdar, A. n.d.. *Coastal Commons. A Glimpse into the Nature and Significance of Coastal Common Spaces and Resources*. Dakshin.

Rudolph, T., Ruckelshaus, M., Swilling, M., Allison, E., Österblom, H., Gelcich, S. and Mbatha, P. 2020. A transition to sustainable ocean governance, *Nature Communications*.

Salleh, A. 2016. Climate, water, and livelihood skills: a post-development reading of the SDGs, *Globalizations*, 952–959.

UN. n.d. *The Sustainable Development Agenda*. Available at: https://www.un.org/sustainabledevelopment/development-agenda-retired/

UNESCO. 2018. *Issues and Trends in Education for Sustainable Development*.

UNESCO. 2021. *Ocean Literacy within the United Nations Ocean Decade of Ocean Science for Sustainable Development*, Paris.

Vân Phạm. 2020, March 14. *VTV - Khẩn cấp về xâm nhập mặn 2020. [VTV – Saline Intrusion Emergency 2020]* [Video]. YouTube. Available at: https://www.youtube.com/watch?v=lou65s3p83E

WWF. 2020, March 03. *Failing SDG14: EU on a Cliff Edge for Ensuring a Sustainable Ocean*. Available at: https://www.wwf.eu/?360550/Failing-SDG14-EU-on-a-cliff-edge-for-ensuring-a-sustainable-ocean

Žalėnienė, I. and Pereira, P. 2021. Higher education for sustainability: a global perspective, *Geography and Sustainability*, 99–106.

ABOUT THE EDITORS

Professor Simon J. Davies has over 40 years of experience in academia with extensive teaching and research attainment. His aquaculture specialisation is in fish nutrition and feed technology, supervising over 40 full-time PhD students and generations of undergraduate and master's students in aquaculture and related marine biosciences. He developed and led master's programs in Marine Biology and Aquaculture at the University of Plymouth (UK) and Animal Nutrition Sciences at Harper Adams University (UK). He is currently an Adjunct Professor at the University of Galway, Ireland, and Emeritus Professor at Harper Adams University. Professor Davies has extensive networks in each continent, speaking by invitation at prestigious international events and scientific meetings, workshops, and symposia.

Paul Robert van der Heijden is a Biochemist, moving into research and development at a large company as a member of techno structure (Mintzberg) and later starting his own business. Inspired by the United Nations (UN) report 'Our Common Future', he undertook an executive MBA education at Wharton Business School USA, at INSEAD France, University Berlin, Germany, and universities in the Netherlands. As an entrepreneur, Mr van der Heijden initiated companies in the People's Republic of China, Indonesia, and the Caribbean and governance became his focus. His relationship with the UN Court of Justice brought him into contact with global stakeholders in sustainable development.

ABOUT THE CONTRIBUTORS

Jeleel Opeyemi Agboola currently working as a Scientist within the Nutrition and Formulation Department at BioMar Global R & D. Jeleel holds a PhD degree in Animal and Aquacultural Science from Norwegian University of Life Science, and dual Master's degree in Animal Science from Wageningen University and Aarhus University. His research focuses on multidisciplinary area of fish nutrition, fish health, gut microbiome, and environmental sustainability. In recognition of his academic and research excellence, he was the winner of Nutreco Young Researchers' Prize 2022, an award aimed at recognising young scientists working at ensuring food security around the world.

Martin J. Baptist is a Dutch Marine Ecologist with an MSc in Environmental Sciences at Wageningen University and a PhD in Civil Engineering at Delft University of Technology. He has a broad understanding of the functioning of marine ecosystems, governing the hydro- and morphodynamics as well as ecosystem management and monitoring, in combination with expertise on coastal safety and dredging. His interest lies in the harmonisation of human activities with natural processes. He applies his knowledge on nature and engineering to complex environmental and spatial issues. Baptist is a renowned expert on the interaction between infrastructural works and ecology and is a leading scientist in the field of Nature-based Solutions.

Matt Elliott Bell was born in Newcastle upon Tyne. He spent his childhood in the Northeast of the United Kingdom and Northumberland coastline, which led to his interest in the ocean. In 2018, he attended Newcastle University's 'Marine Pathways' course at the Dove Marine Laboratory, Cullercoats. This demonstrated how Marine

Science was conducted at the degree level through a scientific research project culminating in presenting his findings to an audience of staff and guests. This was the best way to understand how science and the implementation of sustainability are conducted at the degree level. In 2022, he commenced a placement year supporting the smooth running of Plymouth University Marine Station's HSE commercial diving course; studying aquaculture at The University of Galway; completing polar training with explorer, Jim McNeill and finally, analysing acoustics of invertebrates, fish, and cetaceans at the University of Plymouth.

Vera Cirinà grew up in Europe and had the opportunity to live in three different countries before moving to England for university. Growing up with a diving enthusiastic father and watching Jacques Cousteau and Hans Hass documentaries, she developed a passion for the underwater world since childhood. She always found underwater life a fascinating subject and has been driven towards understanding this topic with a motivation to protect and observe this environment. Her multifaced curiosity led her to the Environmental Science degree because of its multidisciplinary curriculum, addressing current global issues, and seeing them not only from an ecological perspective but also an economic and social one.

Chris de Blok, is an ecologist. He grew up in a small village along the North Sea, in a house that lies 200m from the sea within the Netherlands. Growing up, his life was spent in or around this sea; while driven by pure curiosity, he has constantly been pushing his limits in the marine environment; scuba diving, snorkeling, freediving, and kayaking. In his professional development, he did research in the fields of fisheries, aquaculture, shallow marine ecosystems, and more recently within the deep-sea domain. His institutional connections are with the University of Applied Sciences, van Hall Larenstein, where he does the Bachelor's study of Coastal and Marine Management, and the company Mature-Development BV. As an ecologist with a fearless curiosity, he is constantly drawn to remote and unknown places in hopes of one day understanding the conundrum of species that these ecosystems create and uphold.

Dominic Duncan Mensah is a Fish Biologist from Ghana, currently working as a PhD student at the Norwegian University of Life Sciences, Norway. He completed his Bachelor's in Fisheries and Aquatic Sciences at the University of Cape Coast in Ghana after which he worked for a year as a Teaching and Research Assistant in the same university. He was a recipient of a two-year Erasmus Mundus joint International Master of Science (MSc) in Marine Biological Resources (IMBRSea) scholarship hosted by Ghent University in Belgium with specialisation in Marine Food Production between 2017 and 2019. Following the completion of this master, he enrolled in the University of Life Sciences in Norway for a second master in Feed Manufacturing Technology in Norway between 2019 and 2021. Currently, he works as a PhD student at the same university, characterising the health-modulating effects of microbial ingredients on Atlantic salmon.

Liv Torunn Mydland is a Professor in Fish Nutrition at the Norwegian University of Life Sciences (NMBU). She holds an MSc (Cand. Scient) in Biochemistry and Molecular Biology from the Department of Biochemistry and Molecular Biology, University of Bergen in Norway and a PhD in Animal Nutrition from the Department of Animal and Aquacultural Sciences, NMBU. Her present research focus is to develop and document high-quality novel feed ingredients for fish and farm animals based on renewable natural resources and side streams by use of advanced biotechnology and biorefinery technology.

Margareth Øverland, Professor, has more than 30 years' experience in animal nutrition research, including leading a Centre of Excellence on alternative protein sources and coordinating several projects on sustainable feed development for aquaculture. She is Centre Director of the Innovation Centre Foods of Norway at the Norwegian University of Life Sciences (NMBU). Ongoing research includes developing sustainable protein sources by upcycling underutilised biomass into microbial ingredients with novel biotechnology. Margareth has BSc and MSc degrees in Animal Nutrition from Montana State University, USA, PhD studies at the University of Minnesota, USA, and a PhD in Animal Nutrition from NMBU.

Richard Page is a Human Ecologist and an Innovation Ambassador located at Palau Community College, former Professor of Political Science at Ocean University of Qingdao China and partner of the UN FAO in developing sustainability systems that have global applications.

Caterina Pezzola is a Marine Biologist from Italy. She is currently a Researcher and Scientific Advisor working in the Netherlands. Her focus is on improving the understanding of seaweed ecophysiology to optimise cultivation for commercial purposes. After a BSc in Biology at the Marche Polytechnic University (Ancona, Italy), Caterina obtained a double MSc in Marine Biology and in Biology – Science, Business, and Policy from the University of Groningen (Netherlands). During her studies, she carried out an extra-curricular internship at the University of Tromsø (Norway) on zooplankton movement dynamics in the Arctic. Later, for her Marine Biology Master's thesis, she collaborated with the Norwegian Institute of Marine Research (IMR), assessing the impact of fish aquaculture on benthic communities through image analysis. Within her second Master's, she undertook an internship aimed at assessing the scientific, legal, and economic possibility of sourcing naturally growing, abundant, and non-valorised seaweed stocks in Italy for commercial purposes.

Richard Pompoes was born in Berlin, Germany, and spent his childhood and teenage years growing up in Budapest, Hungary, and Sofia, Bulgaria. He moved to the Netherlands in 2015 to start his education as a water professional at HZ University of Applied Sciences, pursuing a Bachelor's degree in Water Management / Aquatic Eco-Technology. After completing his Bachelor's degree, Richard enrolled at Utrecht University for a Master's degree in Sustainable Development, specialising in International Development, which he completed in 2022. He is currently working as a PhD Researcher at Wageningen University and Research, at the Public Administration and Policy Chair Group.

James Logan Sibley, a Northeastern University alumnus in Cell Biology, 2023, is known for his significant achievements in aquaculture and digital media. His technical background coupled with his passion for science communication and aquaculture has seen his meteoric rise in educational social media. Today, James teaches his growing global audience about the intricacies of modern aquaculture and adjacent biotechnologies. Through professional and personal endeavours, he has made notable contributions to promoting the role of technology and genetic research in sustainable seafood production.

Davy van Doren is a Scientific Researcher with an MSc in Biology and Sustainable Development. His current work focuses on the modelling and assessment of sustainable energy systems, the socio-economic analysis of emerging technology, and the assessment of innovation policy. In addition, he has been involved in multiple projects of MatureDevelopment related to aquatic cultivation systems and biobased remediation solutions. He has also engaged in the nurturing of interdisciplinary learning and sustainable entrepreneurship among young professionals. Furthermore, he has conducted research in relation to the monitoring and preservation of marine ecosystems. Within this context, he has worked as an environmental engineer for the Dutch dredging company Boskalis, where he joined the Building with Nature team to develop an integrative approach towards coastal protection and infrastructure construction.

Printed and bound by CPI Group (UK) Ltd, Croydon, CR0 4YY

08/10/2024

14570957-0001